A IS FOR ATLAS

# A IS FOR ATLAS

## WONDERS OF MAPS AND MAPPING

Megan Barford

NATIONAL
MARITIME MUSEUM
GREENWICH

# INTRODUCTION

**AROUND 1909, EMMY INGEBORG BRUN,** a Danish woman with keen interests in astronomy and politics, painted over the surface of a terrestrial globe. Using blue and cream hues, with black for detail, she carefully depicted a topography of Mars (from a map by American astronomer Percival Lowell), covering the globe's surface with a criss-cross network of 'canals' (image 1). Although it is now recognised that the straight lines of canals were optical illusions, they were the subject of much astronomical debate and publicity around the turn of the twentieth century. These lines, first described as 'channels', and later as 'canals', took on significance as possible evidence of life on Mars and were reported widely as such in the press. While professional astronomers had moved on from serious engagement with the theory of Martian canals by the time Ingeborg Brun painted them onto her globe, they remained an important part of popular discourse on the Solar System, raising questions about the nature of extra-terrestrial life. If the canals existed, who built them and how was their society organised? Were the canals evidence of entrepreneurial genius laid out over a whole planet, or of a cooperative, cosmopolitan society? With her globe (or globes, for she made many) Ingeborg Brun took a position on this matter too. Engraved on a specially made stand were the words 'Free Land, Free Trade, Free Men', a contemporary socialist slogan. A network of canals across the planet implied global cooperation, which allowed precious water to flow to areas which needed it most (from the blue areas on her model to the cream). A second engraved line, 'Thy will be done on Earth as it is in Heaven' from the Lord's Prayer, implores such cooperation in the human world.

   The story of Emmy Ingeborg Brun's Mars globe serves as a good introduction to this book thanks to the different ways it requires us to think about maps and mapping. The globe grabs our attention precisely because it is a surprising representation of a planet mostly known for being red. Here, it is blue, and

[1] *Mars efter Lowell's Glober,* Emmy Ingeborg Brun, 1909

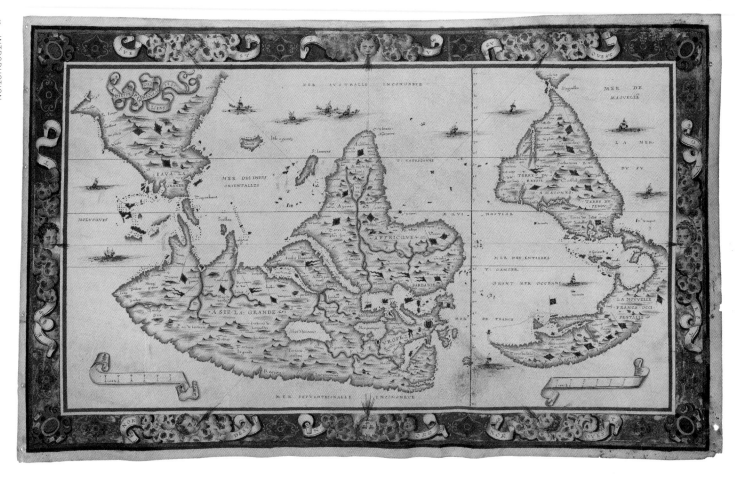

covered in lines that we learn represent canals. But resting secure in the knowledge that we know better now and looking with satisfied superiority at the past does not get us very far. Instead, we must ask questions about the world from which such an object emerged, so that it becomes understandable. It is work historians sometimes describe as 'making the strange familiar'. The opposite, 'making the familiar strange', involves questioning the assumptions and expectations we hold. That might involve asking why on maps of parts of the Earth's surface north tends to be at the top, or why the constellations are grouped in particular ways on star charts. Such shifts in focus are incredibly useful when thinking about maps (image 2). On the one hand, if maps are tools we regularly use, we can feel very comfortable with what they are and their purpose. On the other, it can be easy to misinterpret, misunderstand, and expect something of an object for which it was simply not designed.

Throughout this book, the word 'map' is taken to have two meanings. In the first instance, it is an overarching category that encompasses all 'graphic representations that facilitate a spatial understanding of things, concepts, conditions, processes or events in the human world'[1] and therefore includes objects – such as globes (spherical representations of planets or

[2, above] World map with south at the top, Nicolas Desliens, 1567

[3, opposite] Half of a world map in two hemispheres, showing North and South America, the Pacific and East Asia, with open ocean at the South Pole, Michaele Tramezzino, 1554

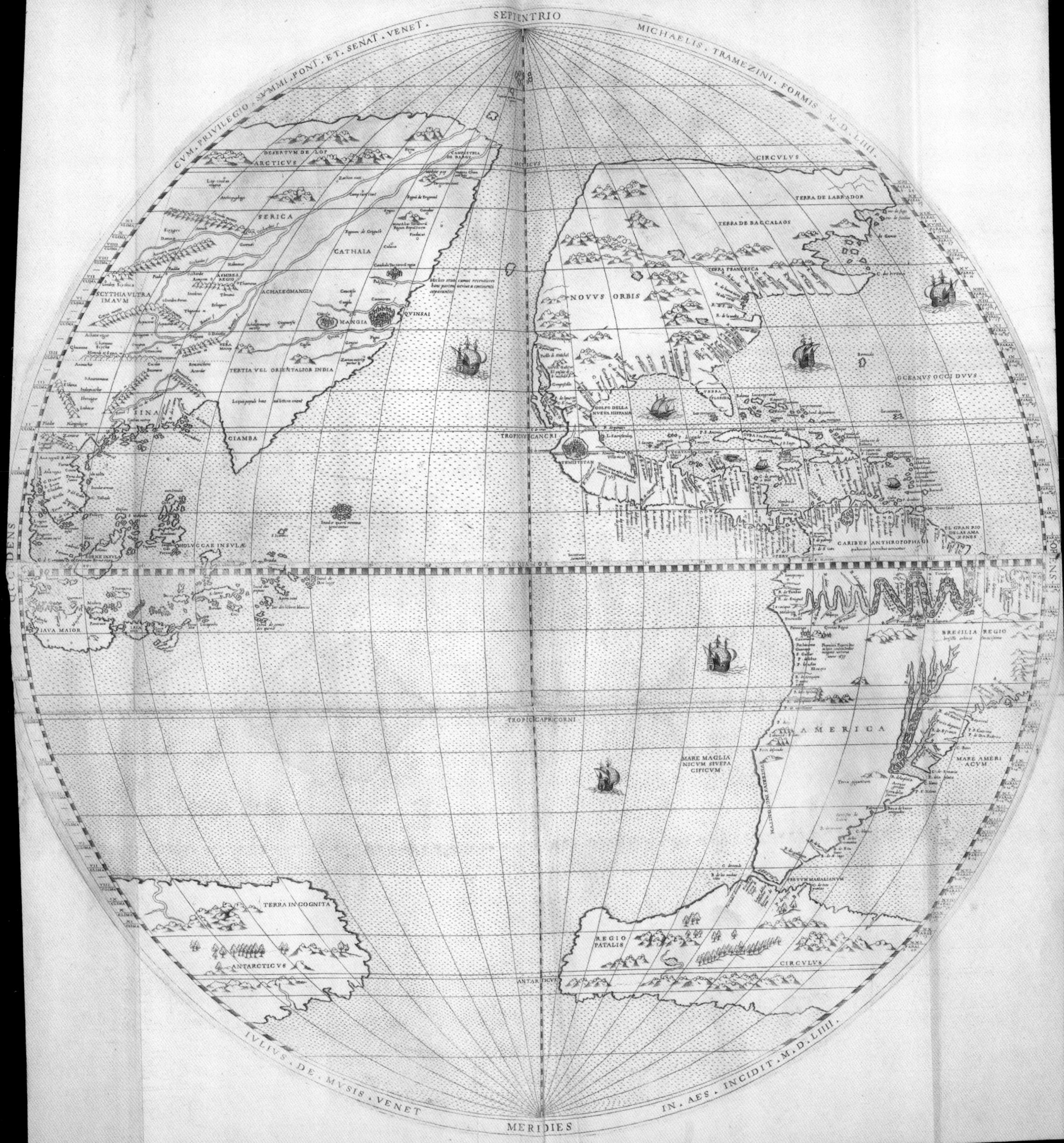
SEPTENTRIO
CVM. PRIVILEGIO. SVMMI. PONT. ET. SENAT. VENET.
MICHAELIS. TRAMEZINI. FORMIS. M. D. LIIII.
IVLIVS. DE. MVSIS. VENET
IN. AES. INCIDIT. M. D. LIIII.
MERIDIES
DESERTVM DE LOP
ARCTICVS
CIRCVLVS
SERICA
CATHAIA
SCYTHIA VLTRA IMAVM
ACHALEGMANGIA
MANGIA
QVINSAI
TERTIA VEL ORIENTALIOR INDIA
CIAMBA
TERRA DE LABRADOR
TERRA DE BACCALAOS
TERRA FRANCESCA
NOVVS ORBIS
OCEANVS OCCIDVVS
GOLFO DELLA NVEVA HISPANIA
TROPICVS CANCRI
TEMISTITAN
MOLVCCAE INSVLAE
IAVA MAIOR
AEQVATOR
CARIBES ANTHROPOPHAGI
EL GRAN RIO DE LAS AMAZONES
BRESILIA REGIO
TROPICVS CAPRICORNI
AMERICA
MARE MAGLIANICVM SIVE PACIFICVM
MARE AMERICVM
FRETVM MAGALIANVM
TERRA INCOGNITA
ANTARCTICVS
REGIO PATALIS
CIRCVLVS
ANTARCTICVS

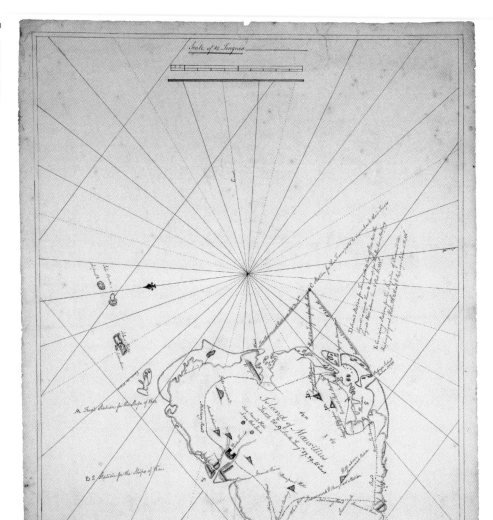

[4] Plan proposing an
English invasion of Mauritius,
James Campbell, 1804

constellations), charts (which tend to be nautical) and plans (very large scale)
– that would be described more specifically in other terms. 'Map' also functions
alongside these descriptors to mean something distinct from 'chart' or 'plan'.
Typically, the word 'map' is used here in the first sense, that is as an overall
category. Throughout, therefore, all charts and globes are maps, but not all
maps are charts or globes.

This book is not a chronological history of maps. Rather, it uses a series of
themes around which we can ask questions of a rich variety of maps, linked to
techniques of making, materials, genres, features and uses or functions. These
themes have been organised alphabetically: E is for Engraving; G is for Globe; N

is for Navigation; P is for Paper; S is for Sea Monster. And under these headings there are stories of cartographic work and skill, of the many minds and hands that went into producing what were often very technically sophisticated as well as artistic objects (image 3). There are stories of imperial mapping (image 4), of the imposition of borders, the exploitation of local knowledge and the writing out of Indigenous presence from a landscape. There are stories too of labour solidarity at sea and of mapping projects designed to reveal the murderous nature of twenty-first-century border policies. Made in specific political and social contexts, all maps present arguments about past, present and future, making assertions about what is important when describing a place. There is, of course, no such thing as a neutral map. All maps, necessarily, are partial, and all maps, therefore, are political.

Technique, material, genre, features and use are, of course, all things which have changed over time. The book explores different methods of production, which often co-existed. The development of copperplate printing following the invention of the printing press, for instance, had a huge impact on map production in Europe but did not replace manuscript techniques. Between 1400 and 1600, the number of maps in circulation in Europe increased dramatically. Indeed, it has been estimated that between 1400 and 1472 just a few thousand maps were made, whereas by 1600 they numbered in their millions.[2] In the nineteenth century, the development of cheaper printing techniques made maps available to much wider audiences (image 5). More recently we have become used to maps not as physical artefacts at all, but as something accessed through a smartphone, which entails a different tactile way of engaging with both map and place. Many of the questions explored in this book, about who makes and controls the data through which huge numbers of people interpret the world, are equally relevant for thinking about newer and older technologies.

Specific genres of map also shift and change; practical features, for instance, become decorative elements. To take one example, a portolan, a major influence on many of the maps in this book, is a type of navigational chart. The *Carte Pisane*, the earliest known portolan and cared for by the Bibliothèque Nationale de France, dates to the thirteenth century. It was designed with the navigational practices of late-medieval Mediterranean mariners in mind, using a system of criss-cross lines (rhumb lines) to help seafarers establish compass bearings. While charts became relatively common navigational aids in regions around the Mediterranean, mariners in northern Europe still largely relied on 'Rutters', textual descriptions of routes between particular ports, until well into the sixteenth century. As long-distance voyages in the years around 1500 spurred the development of overseas colonial and trading empires, sea charts became even more closely linked to displays of wealth and power (image 6). Portolan-style charts began to be produced for decorative purposes, or portolan features were borrowed for maps of a scale too small to ever be navigationally useful (image 7). And while rhumb lines were still added to some navigational charts into the nineteenth century, changing

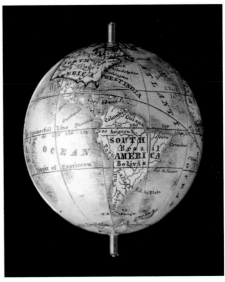

[5] Miniature terrestrial globe and its case, Johann Klinger, 1851, given to a child as a gift by their grandmother in 1870

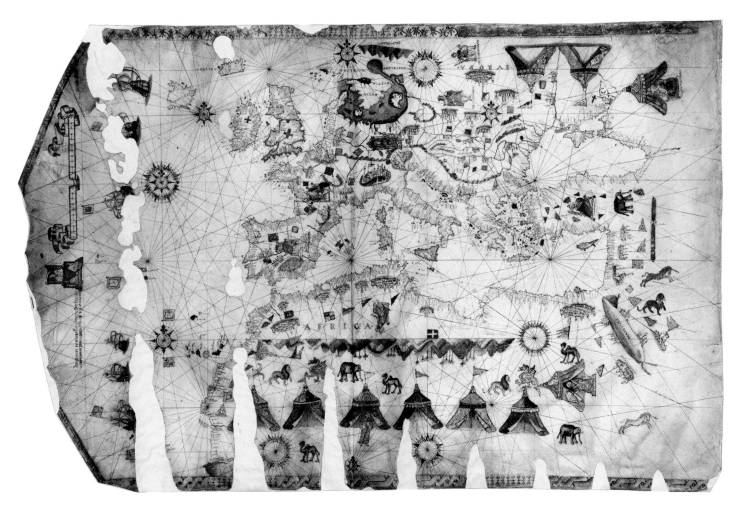

navigational techniques, in particular greater use of astronomical methods, meant that they became less and less important.

Whether used for wayfinding, for display or for teaching, the interpretation and understanding of maps is dependent on the context in which they are made, viewed and used. Indeed, maps and the ways in which they are produced, circulated and consumed, can both conceal and reveal. Made in a variety of contexts and nearly always the work of multiple hands, they can assert particular forms of agency, or the agency of particular individuals or communities, and deny others. In making and in use, maps are part of rich worlds of interpretation and understanding, and this means they have a varied and often fluid status – as artworks, as political instruments, as pragmatic resources, as forms of memory, as tools for planning, as narrative devices. One map tells many stories. Rather than something fixed and finished, individual maps can be reconceived, made and re-made, used and used again, cut up and repurposed (image 8), or thrown away. Or kept – sometimes even in a museum.

*A is for Atlas* draws entirely on the map collection of the National Maritime Museum in Greenwich, UK. As such, while this is not a book solely about sea charts, there is a certain maritime and navigational emphasis to the selection

[6, above] **Portolan chart,** Vesconte Maggiolo, 1548

[7, opposite] *The sea coast from Hormuz to Calicut*, Angelo Freducci, 1555

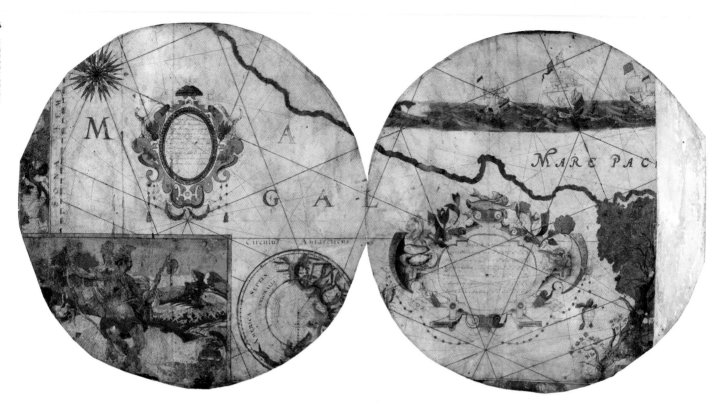

of objects. Indeed, as a journey through the collection, its contents highlight the range of maps that can be consulted in the Museum's stores and library, which are open to the public. And while some of these treasures are illuminated in gold, others are scuffed, stained and scribbled over. They are treasures not because they are particularly shiny (though some are very shiny indeed), but for the way we can use them to understand more about the past and for the questions they provoke about the worlds from which they came.

The map collection of the National Maritime Museum includes around 40,000 objects, among which are loose sheet maps and charts, bound volumes of manuscripts, printed atlases and globes. When the Museum was founded in 1934, it was intended that the collection would show the development of navigational charts in the Western world, demonstrate cartographic developments from the fifteenth century on and illustrate changing European conceptions of global geography. Referring to the acquisition of a sixteenth-century sea chart in 1935 (image 9), Geoffrey Callender, the Museum's first director and the driving force behind the map collection, enthused, 'We, more than any nation, have charted the outer seas, the oceans of today; & it does behove us to illustrate adequately the labour of those who were labouring before we began. Our pictures, even our Models, may leave "NAVIGATORS" cold; but the Zurich Portulan will bring all heirs of Drake and Captain Cook to worship at our shrine.'[3] It was very purposefully an Anglo-centric vision, which positioned English navigators in the present as inheritors of a map history

[8] Circles of vellum cut from a world map by Harmen and Marten Jansz, 1606, possibly for drum tops

[9, opposite above] 'The Zurich Portulan', attributed to Pedro Reinel, around 1535

[10, opposite below] *British Empire Shipping*, The British Admiralty, 1937

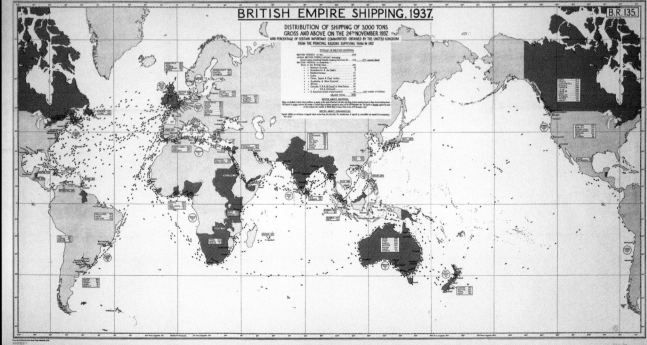

# BRITISH EMPIRE SHIPPING, 1937.

## DISTRIBUTION OF SHIPPING OF 3,000 TONS GROSS AND ABOVE ON THE 24TH NOVEMBER, 1937.

AND PERCENTAGE OF CERTAIN IMPORTANT COMMODITIES OBTAINED BY THE UNITED KINGDOM FROM THE PRINCIPAL REGIONS SUPPLYING THEM IN 1937

B.R. 135.

confined to European and Islamic work and celebrated the British Empire as a source of wealth and global power while glossing over its violence, exploitation and repression.

Today, the Museum seeks to tell different stories. The collection itself has grown and changed since the 1930s and now includes many more objects from the nineteenth, twentieth and twenty-first centuries (image 11). The thematic and material breadth is substantially greater and speaks more to the varied role and status of maps in different contexts. Historians now think more broadly about maps too, making a concerted effort to step away from narratives of progress, in which maps – all judged by the same criteria of accuracy – get better over time. Instead of seeing the world through that

[11] Map of Lesvos and the Aegean Sea, made as a source of information for refugees arriving on the island, Internews, 2015

به جزیره لسبوس خوش آمدید.

در ابتدا باید ثبت نام خود را آن سوی دیگر جزیره ، از طریق مسوولین ذیربط انجام دهید.

۹۰ دقیقه طول می کشد تا از طریق اتوبوس به این منطقه برسید.

افراد سالم و بالغ دو روز نیاز دارند تا این مسیر را به صورت پیاده طی نمایند.

لطفا با همسفران خود باقی بمانید و سوار وسایل نقلیه ي جداگانه نشوید.

همواره مواظب چمدان های خود باشید.

مواظب باشید بهای بیشتری برای کرایه تاکسی و خریدهای کلی نپردازید.

از استراحت و خوابیدن در جاده ها اجتناب نمایید، این عمل بسیار خطرناک است.

Welcome to the Greek island Lesvos.
You must register with the authorities first at the port of Mytilene.
By bus takes about 90 minutes. Walking takes two days.
Please stay with your traveling group, and don't get into separate vehicles.
Keep your luggage with you at all times. Beware of over-pricing.
Do not sleep or rest on the roads.

Καλώς ήρθατε στο Ελληνικό νησί , Λέσβος.
Θα πρέπει πρώτα να πραγματοποιήσετε την εγγραφή σας στο λιμάνι της Μυτιλήνης.
Η διάρκεια για να φτάσετε εκεί με το λεωφορείο , είναι 90 λεπτά.
Είναι 2 μέρες με τα πόδια .
Παρακαλούμε μείνετε με την ομάδα σας, και μην μπαίνετε σε ξεχωριστά αυτοκίνητα.
Κρατήστε τις αποσκευές σας , μαζί σας καθ'όλη την διάρκεια.

particular lens, contemporary accounts of maps and mapping ask more nuanced questions that seek to understand maps in the terms of the communities that made and used them, as well as uncovering those stories that particular maps might suppress.

The production, development and use of maps ultimately reflect our human desire to understand the world around us and to establish our place within it. Studying maps, however, introduces an unavoidable irony. As we explore their variety, it is easy to get lost. The selection of maps assembled here and their organising themes are designed to provide some guidance and direction for the wandering enthusiast, and should help us think about how, why and in what terms we map the world today.

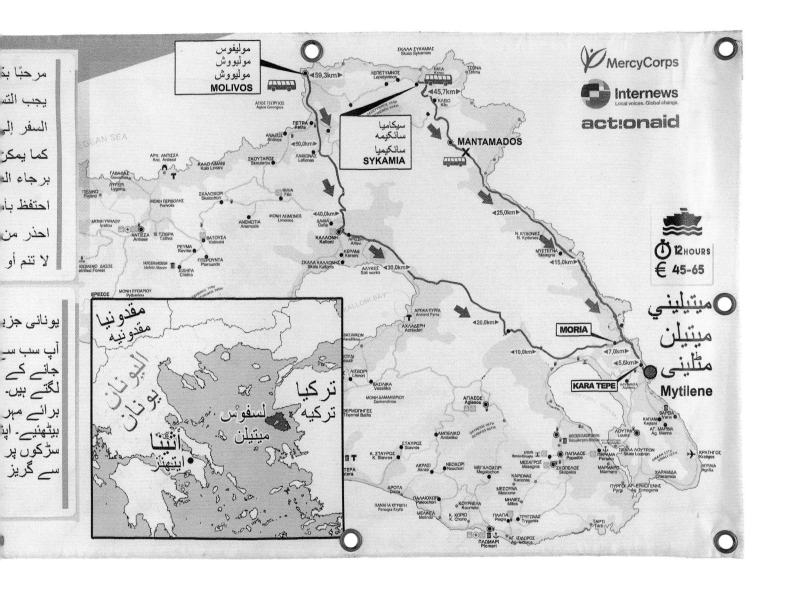

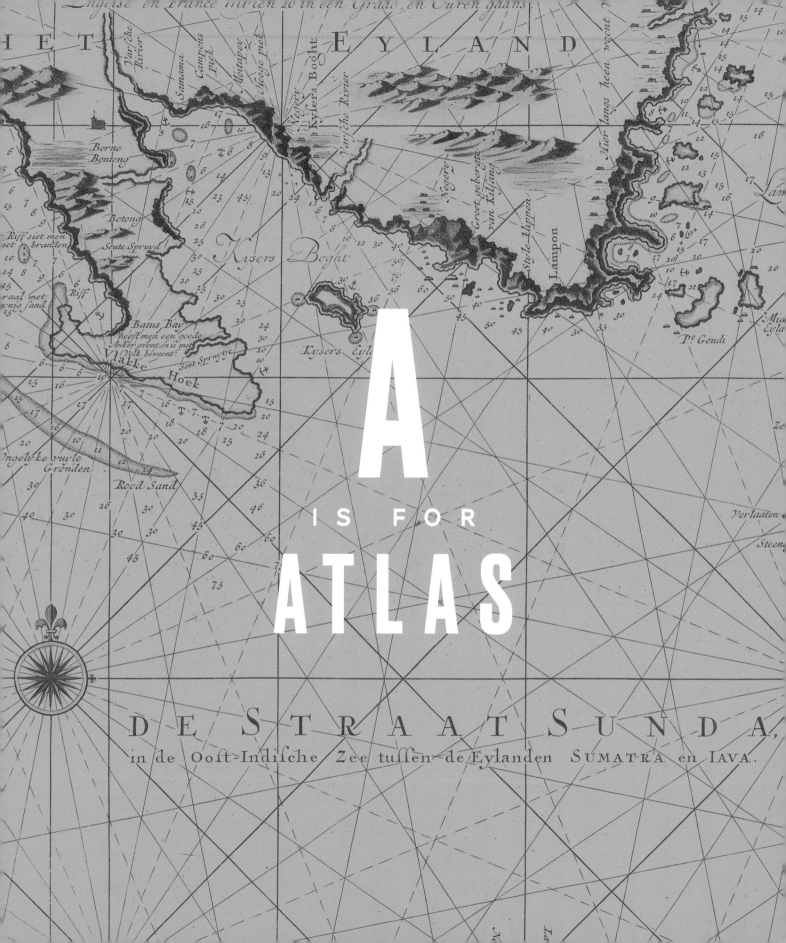

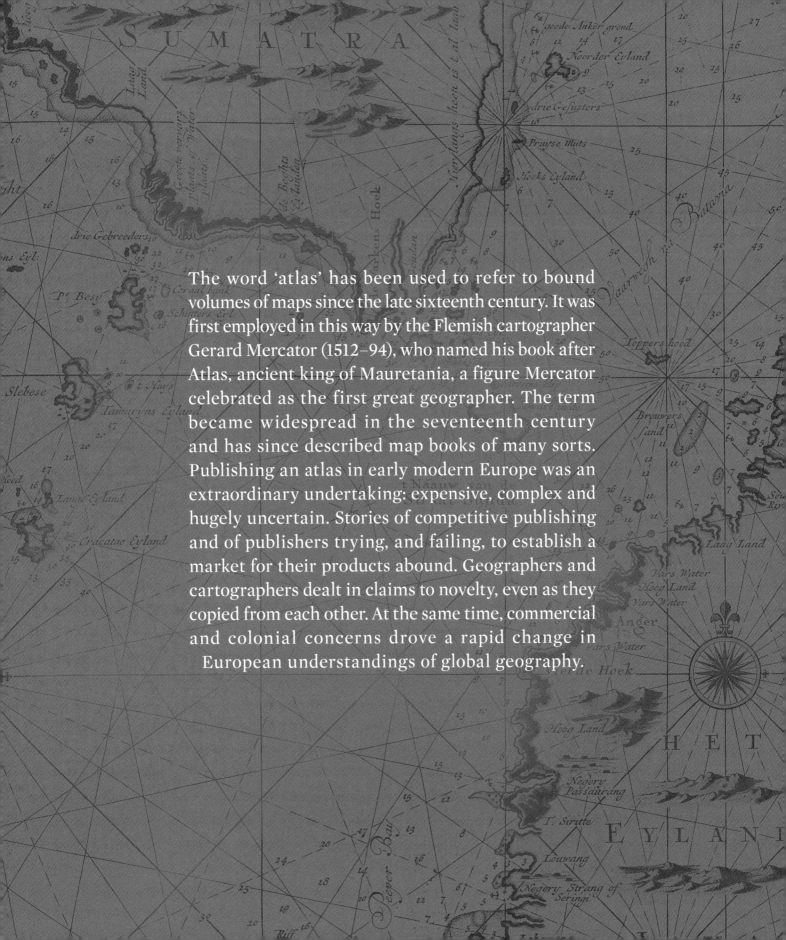

The word 'atlas' has been used to refer to bound volumes of maps since the late sixteenth century. It was first employed in this way by the Flemish cartographer Gerard Mercator (1512–94), who named his book after Atlas, ancient king of Mauretania, a figure Mercator celebrated as the first great geographer. The term became widespread in the seventeenth century and has since described map books of many sorts. Publishing an atlas in early modern Europe was an extraordinary undertaking: expensive, complex and hugely uncertain. Stories of competitive publishing and of publishers trying, and failing, to establish a market for their products abound. Geographers and cartographers dealt in claims to novelty, even as they copied from each other. At the same time, commercial and colonial concerns drove a rapid change in European understandings of global geography.

*Theatrum Orbis Terrarum*, or 'Theatre of the World', by Abraham Ortelius (1527–98) is often described as the first modern atlas. The volume presented 53 maps of uniform size, which were ordered by region, to give a picture of world geography in a way that had never been done before. Each map had text printed on the reverse of the sheet because, Ortelius explained, 'we thought it would not please the reader or spectator to see the back of the map completely empty.' Ortelius himself travelled extensively in the Low Countries, France, Germany, Italy, England and Ireland. He also corresponded widely with scholars across Europe, from whom he sought detailed geographical knowledge, and who he acknowledged in a 'list of authors' in his work. First published in Antwerp in May 1570, *Theatrum* was a triumph. By 1612 it had been published in 31 editions in 7 languages (Latin, Dutch, French, German, English, Spanish and Italian). New editions were expanded, with volumes swelling to over 150 maps, and Ortelius's list of contributors grew from 87 people in 1570 to 183 in 1603.

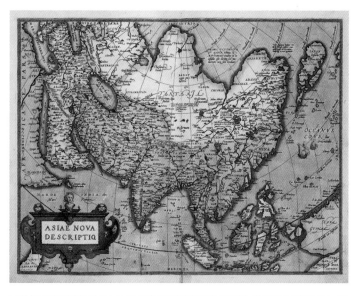

'Typus Orbis Terrarum' (main image), 'Asiae Nova Descriptio' (left, top), 'Hollandiae antiquorum Catthorum sedis nova descriptio' (left, bottom), *Theatrum Orbis Terrarum*, Abraham Ortelius, 1573

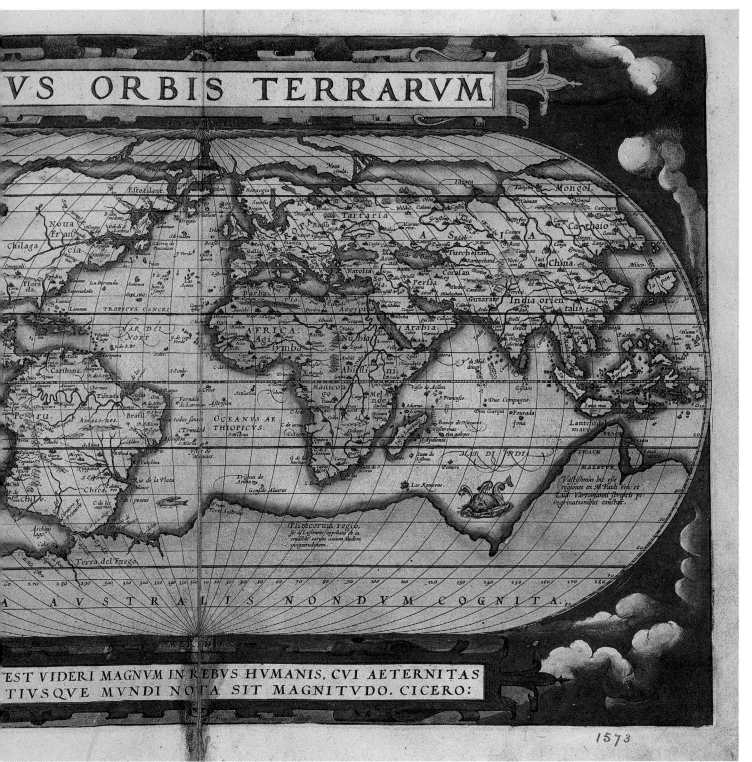

VS ORBIS TERRARVM.

ESST VIDERI MAGNVM IN REBVS HVMANIS, CVI AETERNITAS
TIVSQVE MVNDI NOTA SIT MAGNITVDO. CICERO:

1573

Joan Blaeu's *Grand Atlas* was renowned for its beautiful, typed lettering and finely executed maps. Most of all, it was famous because it was very, very big. And to Dutch map publishers of the mid-seventeenth century, size really did matter. Indeed, it was a rivalry with another map publisher, Janssonius, that spurred the Blaeu publishing house, first under Willem Blaeu and then his son Joan, to produce larger works containing ever more maps. Thanks to his position as cartographer to the Dutch East India Company, Blaeu could afford to compete. His *Grand Atlas*, originally published in 1662, was, at the time, the largest and grandest atlas ever produced. The first edition, in Latin, comprised 11 volumes, each around 58 cm high; the French edition, first published in 1663, came in 12 volumes. The work was intended for a very wealthy audience: a coloured copy of the French edition would cost 450 florins, about the same as a master carpenter would earn in a whole year. It was such a prestigious work that the *Atlas* was frequently given as a gift on behalf of the Dutch Republic: to Admiral Michiel de Ruyter, for instance, after a victory in the Second Anglo-Dutch War (1665–67); to the Ottoman Sultan Mehmed IV, as a diplomatic gift in 1668. This example is a copy of the second French edition, the last edition published by Blaeu when it was printed in 1667.

'Nova et Accuratissima totius Terrarum Orbis Tabula' (main image), the twelve volumes of *Le Grand Atlas*, Joan Blaeu, 1667

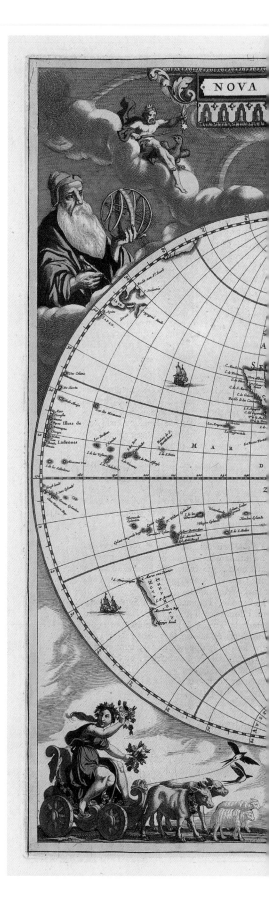

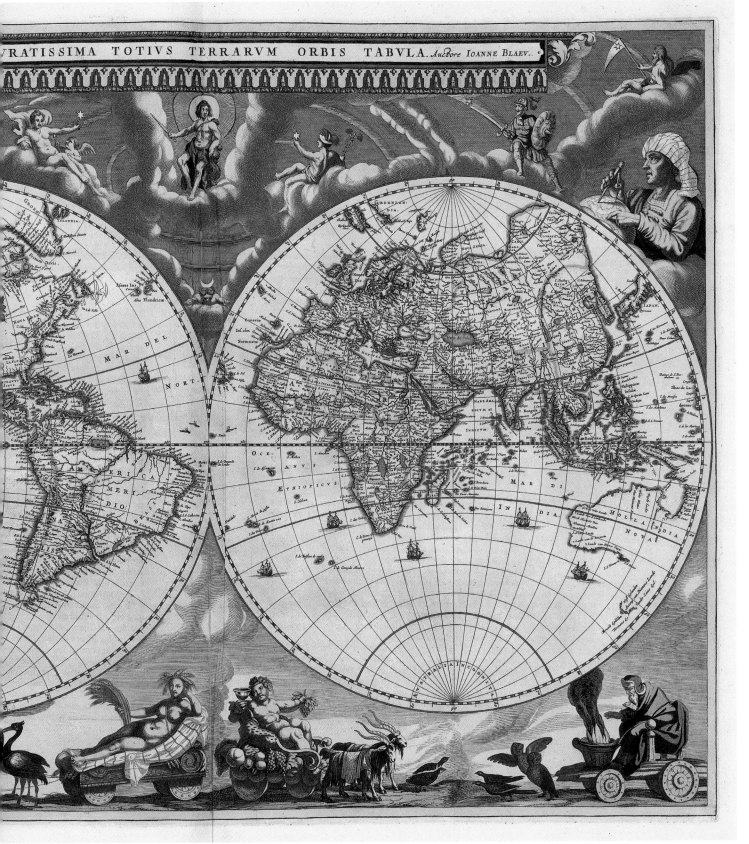

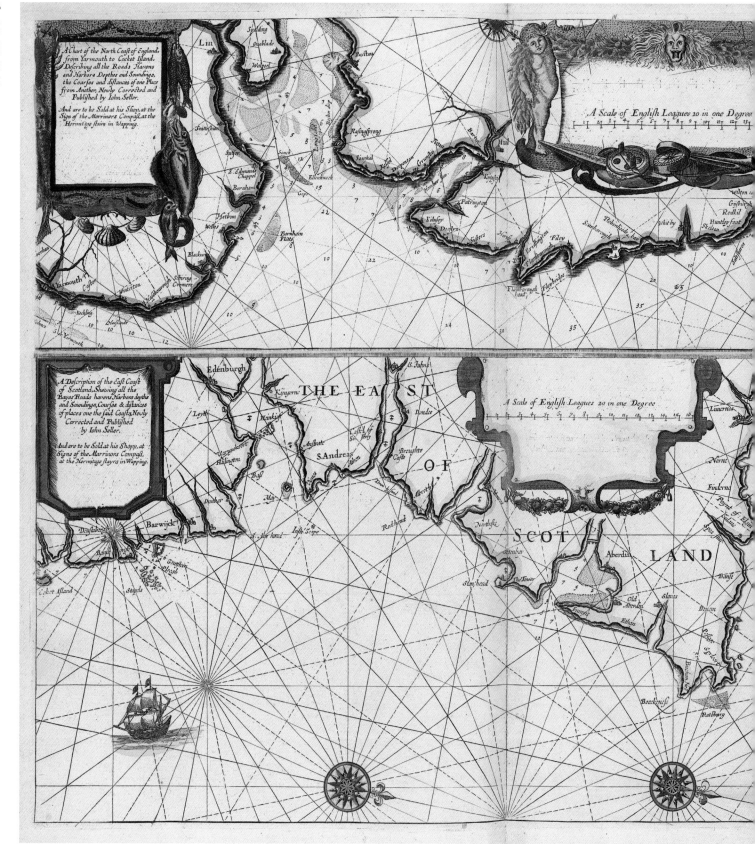

A Chart of the North Coast of England,
from Yarmouth to Cocket Island,
Describing all the Roads Havens
and Harbors Depthes and Soundings,
the Coarses and distances of one Place
from Another, Newly Corrected and
Published by John Seller.

And are to be Sold at his Shop, at the
Sign of the Marriners Compass, at the
Hermitage staira in Wapping.

A Scale of English Leagues 20 in one Degree

A Description of the East Coast
of Scotland, Sheuing all the
Bayes Roads havens Harbons depths
and Soundings, Courses & distances
of places one the said Coast, Newly
Corrected and Published
by John Seller.

And are to be Sold at his Shopp, at
Signe of the Marriuors Compass,
at the Harmitage stayrs in Wapping.

THE EAST

OF

SCOT LAND

A Scale of English Leagues 20 in one Degree

John Seller's *The English Pilot*, first published in 1671, was celebrated as the first English sea atlas. It was presented as a triumph of England's maritime activity, 'furnished with New and Exact Draughts, Charts, and Descriptions … from the latest and best Discoveries … of our *English* Nation'. Despite its claims to newness, and its patriotic overtones, Seller's publication included many charts made from old plates previously used to print Dutch atlases. As the diarist Samuel Pepys described, Seller 'bought the old worn Dutch copper plates for old copper, and had them refreshed in several places'.[1] Although the plates were altered, evidence of the changes appears on the printed sheets. On this map of the east coast of Scotland, you can see where Dutch place names have not been erased, or have been erased, but not replaced with their English equivalents (detail, below). Re-using plates was, however, standard practice. In the seventeenth century and beyond, and particularly in the expensive world of atlas publishing, the history of European cartography is one of copying and re-use. Seller's *English Pilot* simply provides us with a good example of what was in fact a common occurrence.

'A Chart on the North Coast of England' and 'A Description of the East Coast of Scotland', *The English Pilot: Describing the Northern and Southern Navigation*, John Seller, 1671

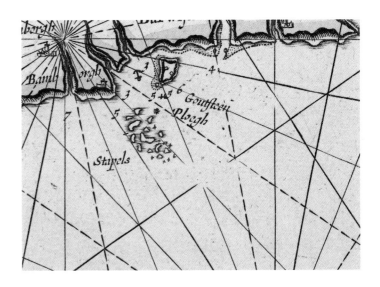

*De Nieuwe Groote Lichtende Zee-Fakkel*, or 'The Great New Shining Sea Torch', originally published in five volumes in the 1680s, was one of the most important sea atlases of the seventeenth century. The sixth volume, pictured here, was not published until 60 years later. Depicting coastlines from the Cape of Good Hope, at the southern tip of Africa, to Japan, via the Indian Ocean and the South China Sea, the information contained was judged to be sufficiently politically sensitive that the charts were not printed but copied only by hand in an attempt to limit their circulation until they were published in 1753. Navigational knowledge of these waters was so important that the charts and sailing directions supplied to Dutch East India Company pilots were closely guarded. By the mid-eighteenth century, however, the situation had changed. As English and French charts of Asian waters increasingly appeared in print, the need for secrecy diminished. The commercial opportunity of publishing the entire set of Dutch charts became clear to the then official supplier of charts to the Dutch East India Company, the House of van Keulen, and this sixth volume of the *Zee-Fakkel* was released. With its fine colouring, this is a presentation copy rather than a navigational one. The chart reproduced here shows the Sunda Strait, between the islands of Java and Sumatra, which was the route taken by Dutch East India Company ships to East Asia.

'De Straat Sunda in de Oost-Indische Zee',
*De Nieuwe Groote Lichtende Zee-Fakkel*,
Johannes van Keulen, 1753

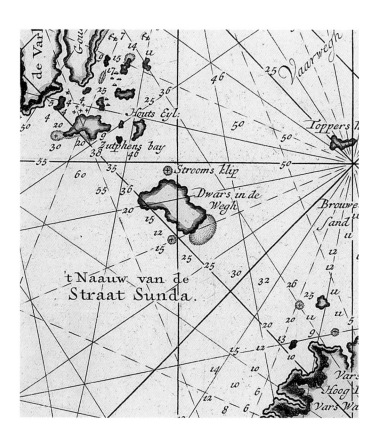

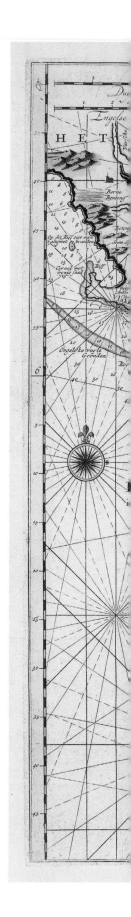

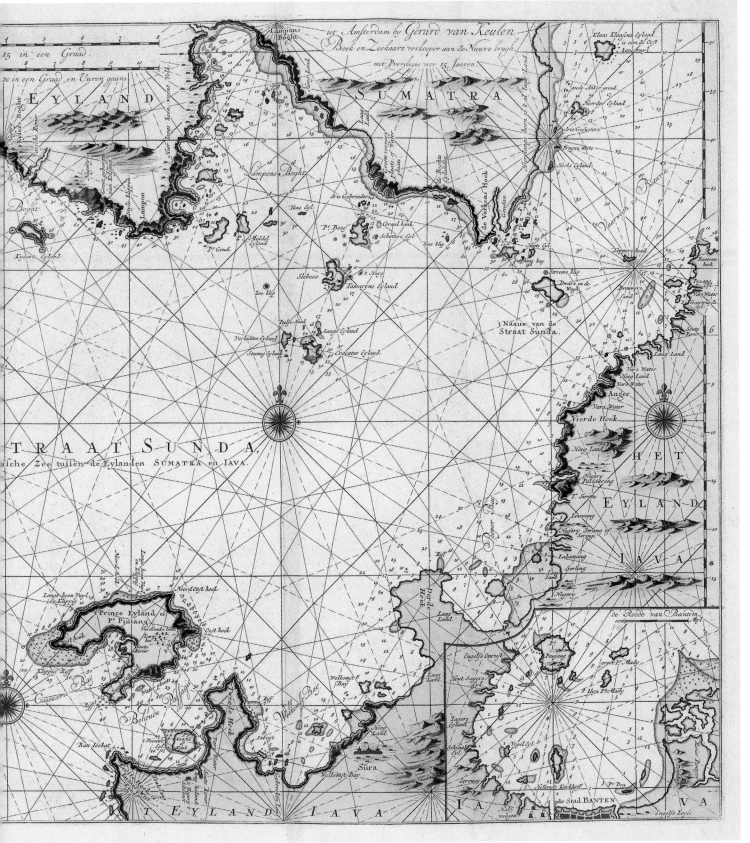

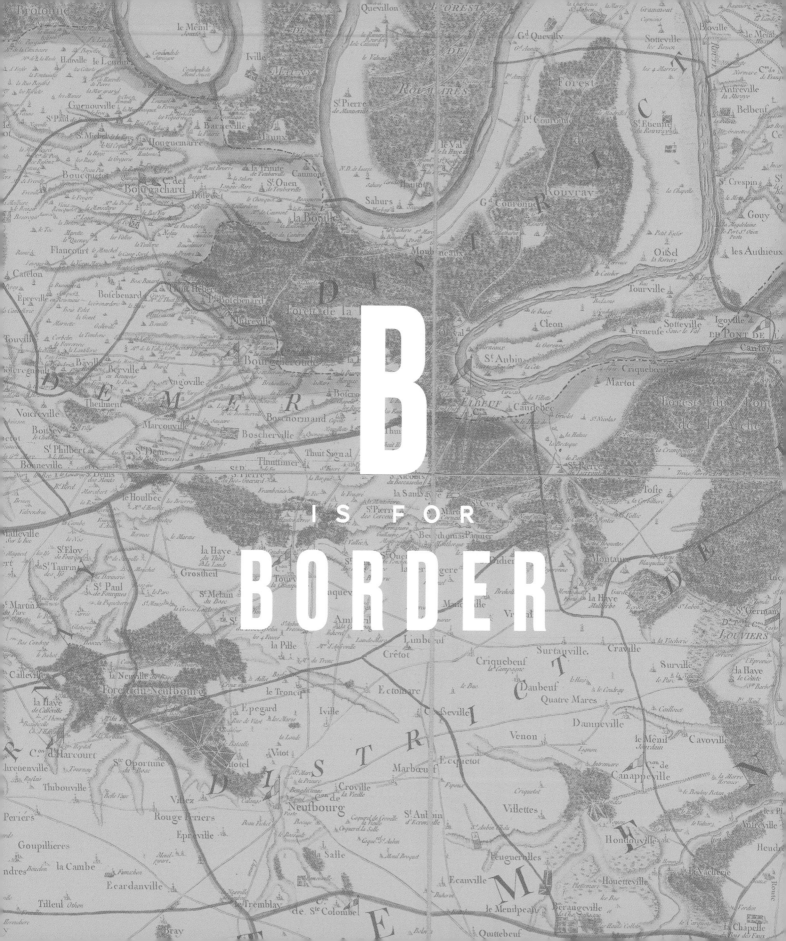

B

IS FOR

BORDER

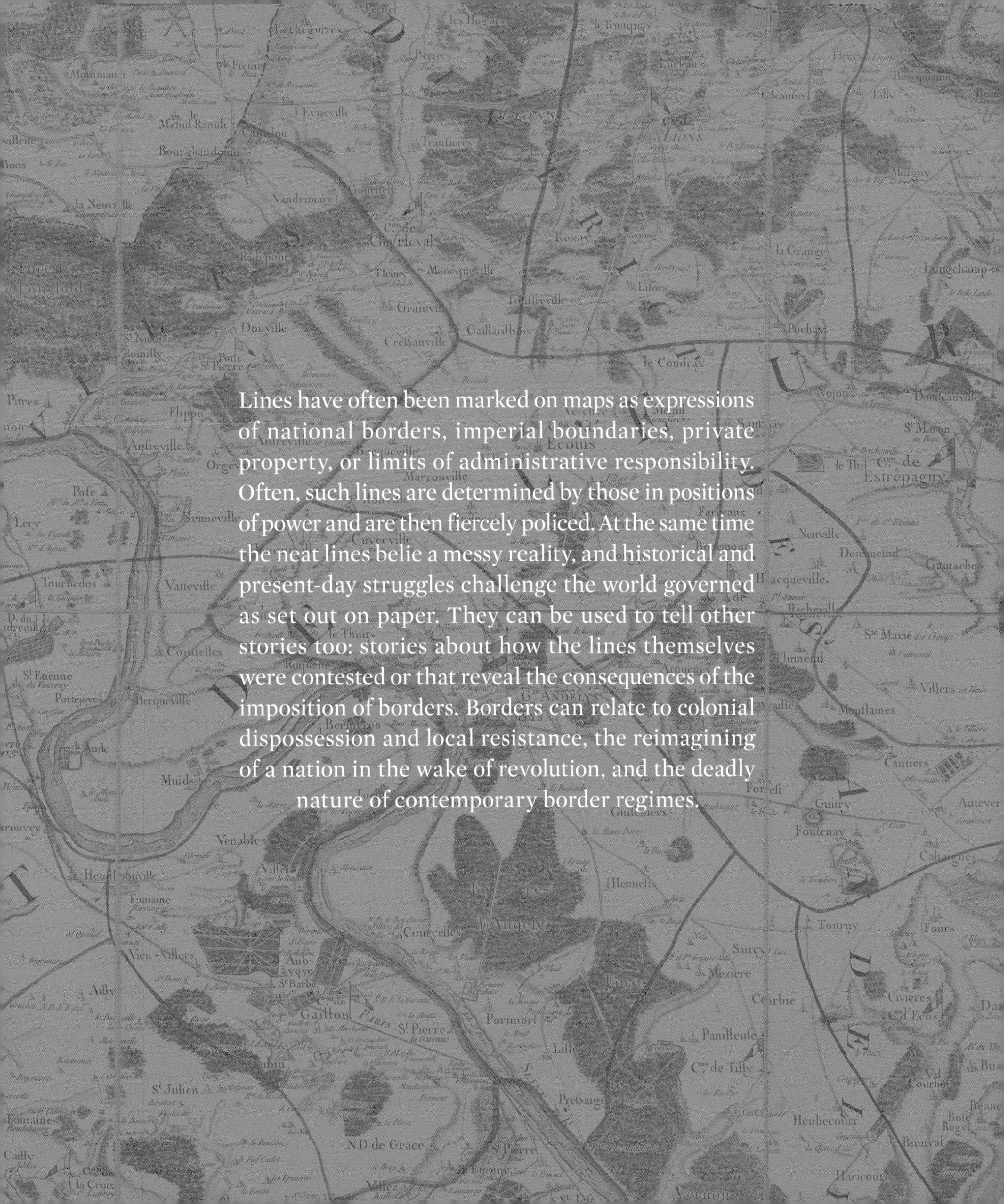

Lines have often been marked on maps as expressions of national borders, imperial boundaries, private property, or limits of administrative responsibility. Often, such lines are determined by those in positions of power and are then fiercely policed. At the same time the neat lines belie a messy reality, and historical and present-day struggles challenge the world governed as set out on paper. They can be used to tell other stories too: stories about how the lines themselves were contested or that reveal the consequences of the imposition of borders. Borders can relate to colonial dispossession and local resistance, the reimagining of a nation in the wake of revolution, and the deadly nature of contemporary border regimes.

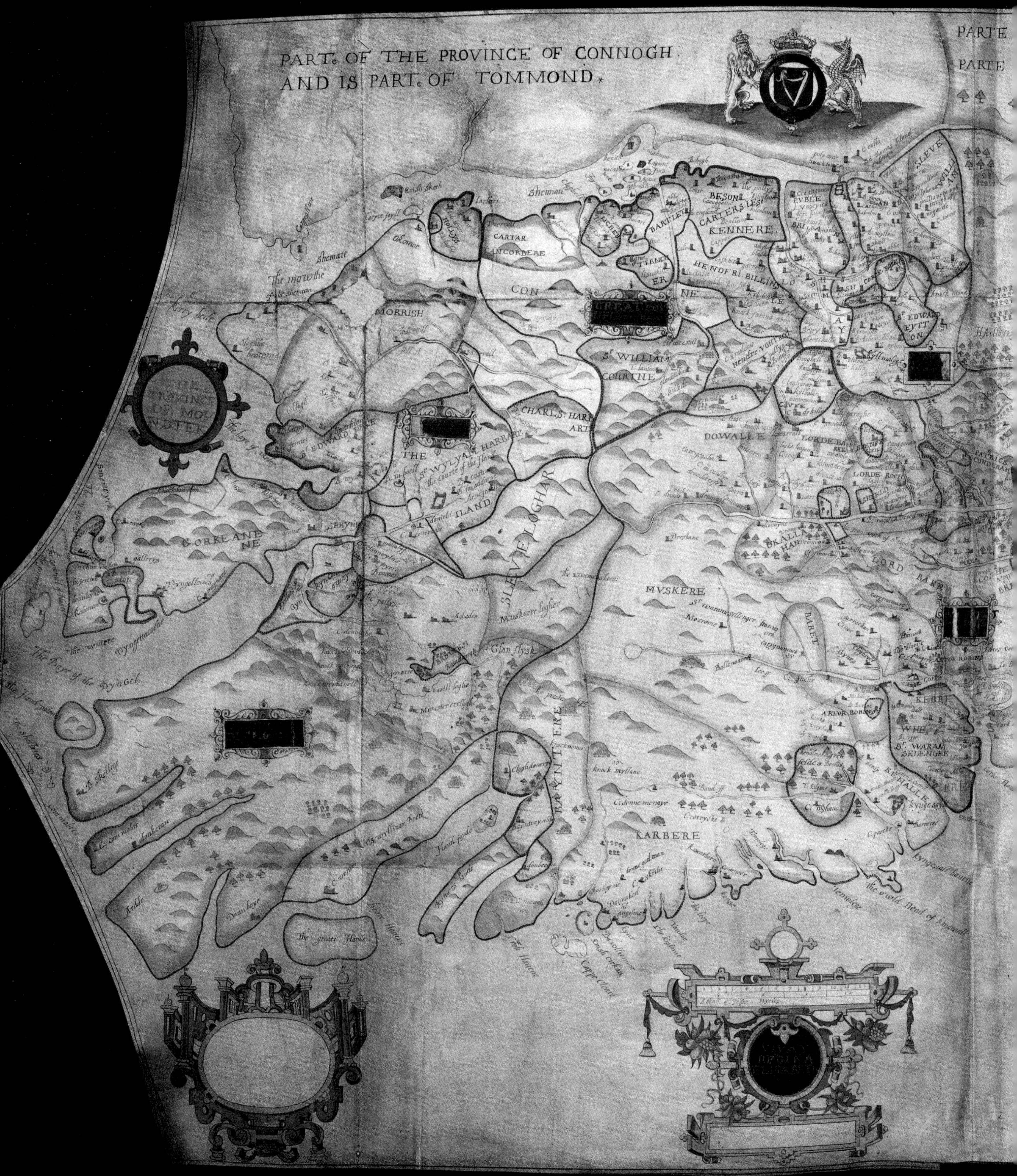

PART. OF THE PROVINCE OF CONNOGH AND IS PART. OF TOMMOND
PARTE
PARTE
BESON
CARTERS LE
KENNERE
CARTAR
CON
MORRISH
St WILLIAM COURTNE
CHARLS HARB ART
THE
St WYLYAM HARBART
ILAND
GORKEANE NE
SLEVE LOGHAR
DOWALLE
LORDE
MVSKERE
LORD BARR
KARBERE
ARTOR ROBINS
St WARAM SELENGER
KENALLE
SLEVE
EVBLE
St EDWARD FYTT ON
The mowthe of the Shennan
The Dyngel
The greate Blaskie

In the latter half of the sixteenth century, a series of rebellions against English rule in the south of Ireland led to the loss of tens of thousands of lives, largely as a result of a scorched-earth policy instigated by the English and their local allies. The English crown confiscated some 500,000 acres of land during this period. The plan was to repopulate the area with English and Welsh landowners and tenants and establish a secure English base in the region. This map of the province of Munster was made as part of this process of plantation, or colonisation by settlement. A commission for 'dividing and bounding [the confiscated lands] into seignories [territories over which a lord has jurisdiction]' led to an extensive survey of the province and a scheme to parcel out land into English-style counties was enacted. Dividing Munster involved imposing a familiar English system of private and inheritable land tenure, which was also underpinned by an ideological assumption that linked such an arrangement to civility and superiority. The drawn borders, which supported neat administrative plans, were, however, never mirrored by reality and the colonial efforts were always challenged. Low-level resistance to the settlers continued through the 1580s and 90s and developed into full rebellion in 1598, which drove out many of the colonists for a time.[2]

*The Province of Munster*,
Francis Jobson, 1589

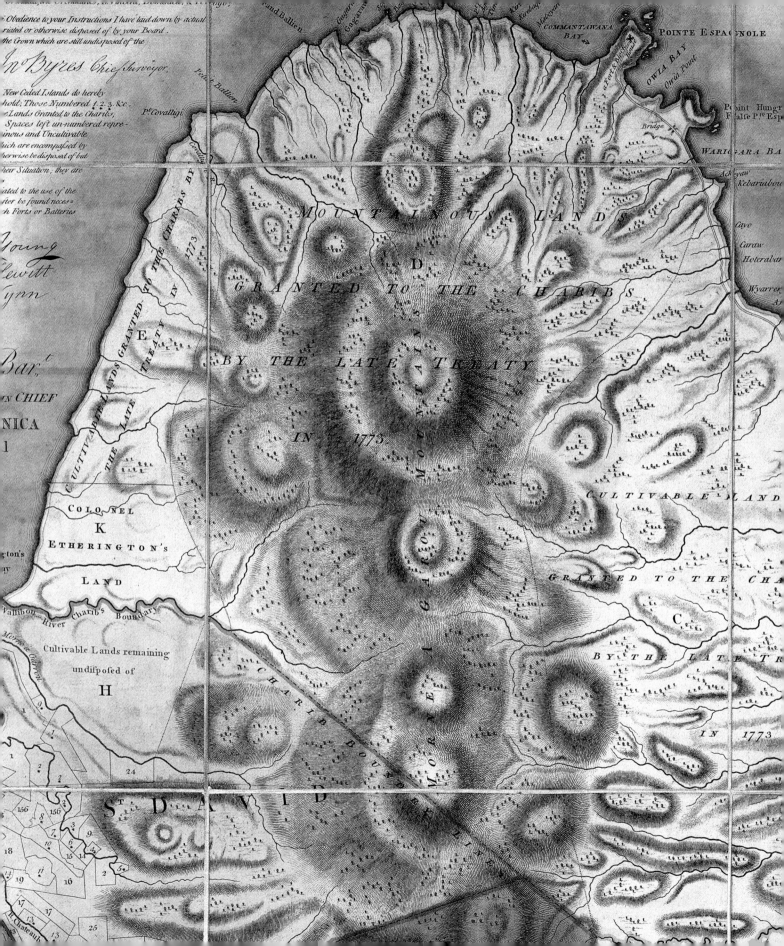

In Obedience to your Instructions I have laid down by actual ... riated or otherwise disposed of by your Board . the Crown which are still undisposed of the

Jn Byres Chief Surveyor.

New Ceded Islands do hereby
hold; Those Numbered 1.2.3.&c .
e Lands Granted to the Charibs,
Spaces left un-numbered repre-
inous and Uncultivable
uch are encompassed by
therwise be disposed of but
heir Situation; they are
iated to the use of the
ter be found neces=
ch Forts or Batteries

Young
Hewitt
ynn

Bart

n CHIEF

NICA

COMMANTAWANA
BAY
POINTE ESPAGNOLE

a Fort & block House
Owia Bay
Owia Point

Point Hungr
Falfe Pt. Esp

Bridge
WARIOGARA BA

Achyau
Kebariabou

Cavo
Caraw
Hoterabar

Wyarar

MOUNTAINOUS LANDS

D

GRANTED TO THE CHARIBS

BY THE LATE TREATY

IN 1773.

CULTIVABLE LAND

GRANTED TO THE CHA

C

BY THE LATE TR

IN 1773

Pedr a Batlien

Pt Covalligi

Covalligi R.

GRANTED TO THE CHARIBS BY 1773

CULTIVABLE LANDS GRANTED TO THE CHARIBS BY THE LATE TREATY

E

COLONEL

K

ETHERINGTON'S

LAND

ton's

y

Vallibou River Charibs Boundary.

Morne a Garou

Cultivable Lands remaining
undisposed of

H

CHARIB BOUNDARY

MORNE A GAROU

MOUNTAINS

1
9.
1
2
24
1

156
156
3
9
8
6
5
7
10
15
14
4
18
14
19
16
2
3
17
17
R. Chateau
13
13
25

S D AVID

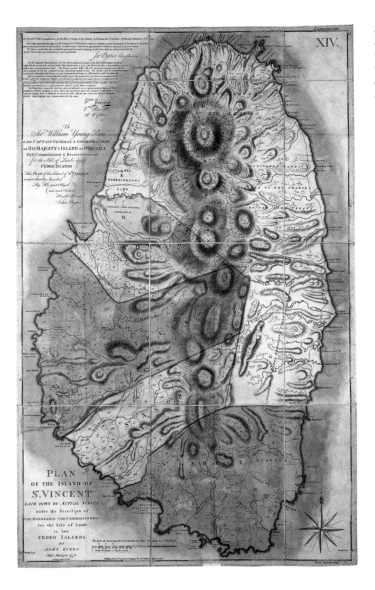

*Plan of the Island of St Vincent laid down by actual survey under the direction of the Honorable the Commissioners for the Sale of Lands in the Ceded Islands*, John Byres, 1776

This map shows one part of a long history of Indigenous Kalinago and Garifuna opposition to European colonisation on the Caribbean island of St Vincent. The island was never colonised by the Spanish and was declared 'neutral' in the 1600s by France and Britain because of the strength of Indigenous resistance. However, due to informal French settlement in the south of the island, St Vincent was considered French territory in the 1763 Treaty of Paris that ended the Seven Years' War. The treaty ceded the area to Britain, in violation of previous agreements. Because of the huge profits that could be made from the labour of enslaved people on sugar plantations in the Caribbean, it was British policy to turn the island into private property for the cultivation of sugar as quickly as possible. A survey was commissioned to divide and sell or lease the land of this new colonial possession. The numbered blocks in the southern part of this plan demarcate the portions of land auctioned in London to those seeking fortunes in the slave and sugar economy of the Caribbean. But the northern part of the island, coloured yellow, is labelled '... granted to the Charibs by the late treaty in 1773'. The border on the map is testament to Garifuna resistance to colonial dispossession. Despite the formal protection of the treaty, the Garifuna faced British encroachment into their territory, which they successfully resisted at first. However, British military intervention intensified in the 1790s and the St Vincent Garifuna people were deported by the British, first to the small island of Belliceaux and then to Roatán off the coast of Honduras. Over half of the almost 5,000 people deported died of disease and starvation on these islands. The Garifuna people continue to live in Belize, Honduras and Guatemala, and once more on St Vincent, as well as in diaspora communities further afield.[3]

The 1789 French Revolution brought sweeping changes to France. The desire to create a blank slate, free from any vestige of the Ancien Régime, involved not only abolishing absolutist rule but also changing the calendar and re-forming national space. In 1790, in place of confusing, sometimes overlapping, sometimes defunct regional administration, the country was divided into 83 departments, themselves split into districts and then communes. This map, one sheet of a huge map of France (the *Carte Géométrique de la France*), has had the borders of the new administrative areas drawn across it in yellow, blue and red. The map is the result of the first systematic survey of France and the work of four generations of the Cassini family – all prominent astronomers. If all the 182 sheets of the map were joined together, it would be about 12 metres high and 11 metres wide. The national military office, the Dépôt de la Guerre, took over the publication of the map in 1793. Through this, a project that was originally commissioned by the King to better understand the territory he ruled became a map to help reimagine the French nation. What is more, the new borders drawn over the top of this map helped demonstrate even more explicitly the new administrative realities of the Republic, intended to break with the provincial jurisdictions that existed before the Revolution.

'Rouen/Départément de l'Eure', *Carte Géométrique de la France*, César-François Cassini, published by the Dépôt de la Guerre, annotated after 1790

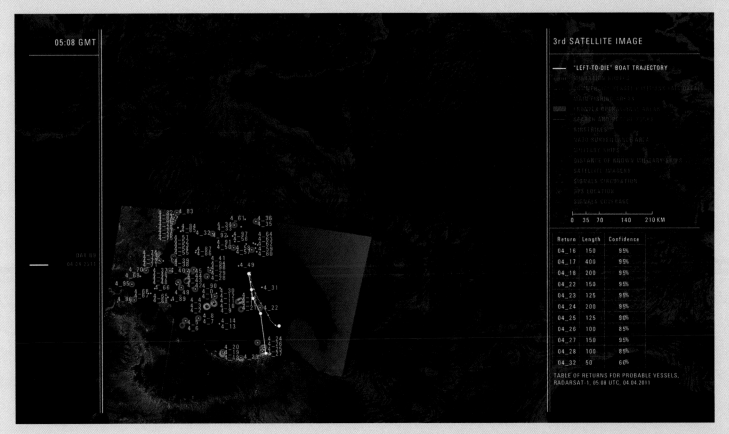

*Liquid Traces: The Left to Die Boat Case*, Forensic Oceanography, 2014

Forensic Oceanography researchers Lorenzo Pezzani and Charles Heller make maps not so much to show borders, but to show their consequences. This map, part of the video work *Liquid Traces: The Left to Die Boat Case*, shows the route of a small boat, which set off from the coast of Libya carrying 72 people during NATO's 2011 military intervention there. The boat's engine failed. Although those on board had repeated interaction with both a military helicopter and vessel, the boat was left to drift for 14 days in some of the most monitored waters on Earth, leading to the death of 63 of its passengers. *Liquid Traces* examines how the intersecting jurisdictions in the Mediterranean allowed states to assume and step back from different responsibilities. Specifically, it interrogates policies and practices of abandonment of migrant vessels in distress and the deadly consequences thereof. Using radar, satellite and vessel-tracking data, alongside individual testimony, Forensic Oceanography describes its work as '[contesting] both the violence of borders, and the regime of (in)visibility on which that violence is founded',[4] referring specifically to the way in which deaths at sea can be easily attributed to the natural environment rather than state policy. This research has formed the basis of legal cases against several European states.

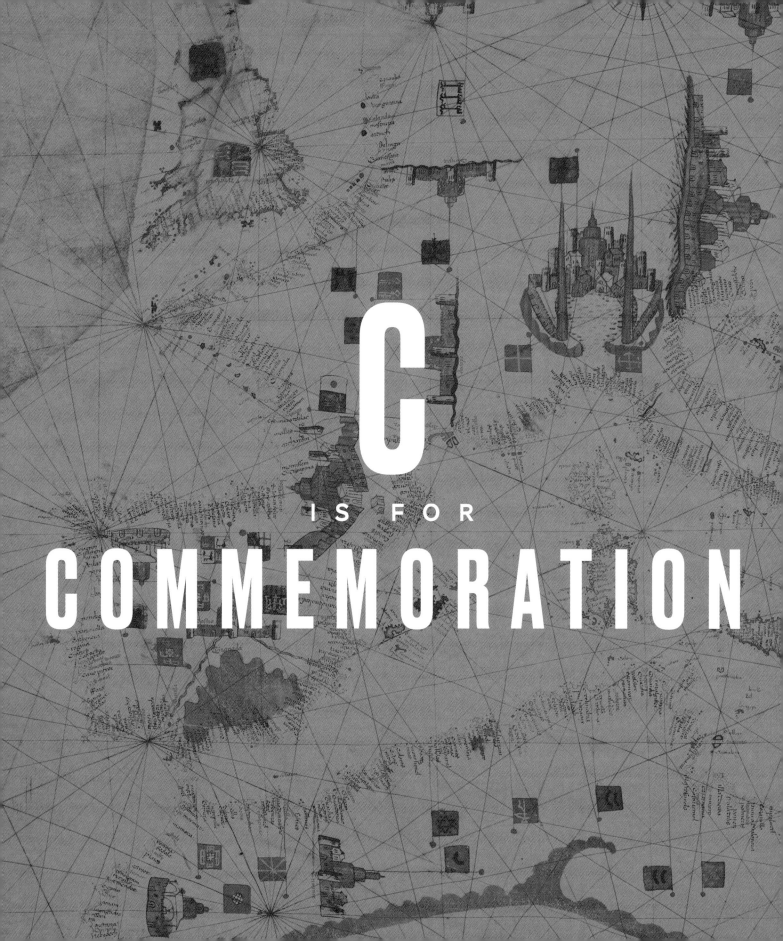

# C
IS FOR
COMMEMORATION

Acts of collective remembrance of particular people or events demonstrate something of the values of a society or community at the moment of commemoration, whether one or one hundred years after the event. Many explicitly commemorative maps relate to success in battle, or to long-distance European voyages which often began or consolidated imperial projects. However, those are not the only things memorialised through maps. Two objects here celebrate the naval victories that for so long epitomised Britain's rise as a maritime power: the defeat of the Spanish Armada in 1588 and the Battle of Trafalgar in 1805. A third recalls one of the most important events in Jewish, Christian and Islamic history. The final object is explicit in its assertion that the process of memorialisation and remembrance is, in fact, about our contemporary attitude to history and lives today.

With its network of criss-crossing 'rhumb' lines, vignettes of port cities, scoop-shaped bays and coloured islands, this is an example of the portolan chart, a navigational document that emerged in the Mediterranean world in the late thirteenth century. It was made in 1456 by cartographers Jacopo Bertran and Berenguer Ripol, who were part of a group of chart-makers – mainly of Jewish origin – based in Barcelona and the Balearic Islands. On this chart the Red Sea is coloured in red (see detail), as was typical for maps of this era. At its northern end a small body of water has been split off from the sea by a thin line that connects the land on either side. Such a marker was not unusual on portolan charts and indicates the passage through the Red Sea by which the Israelites passed from slavery in Egypt to freedom in the Promised Land, as described in the Book of Exodus, part of the Hebrew scriptures. The presence on a map of one of the major events of, in this case, Jewish history demonstrates how geography was infused with meaning in the Renaissance world. A document designed with navigational purpose in mind, the chart depicts the parted waters of the Red Sea not as a historical curiosity but as a way of bringing it into the present for its users, as something continually lived as well as something already happened.

Portolan chart of the
Mediterranean, Jacopo Bertran
and Berenguer Ripol, 1456

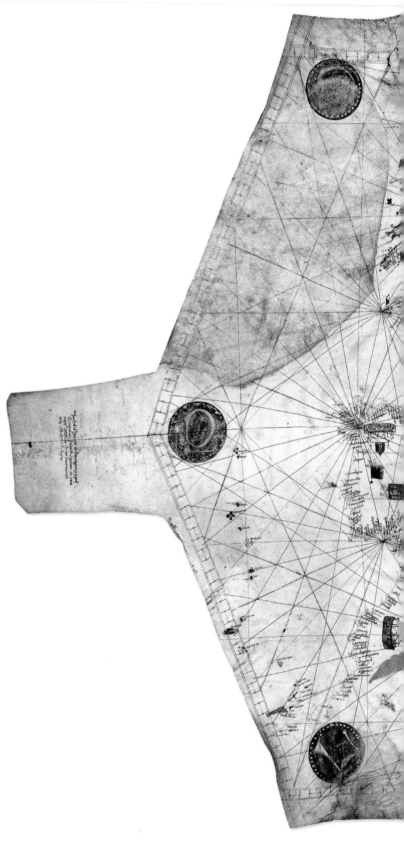

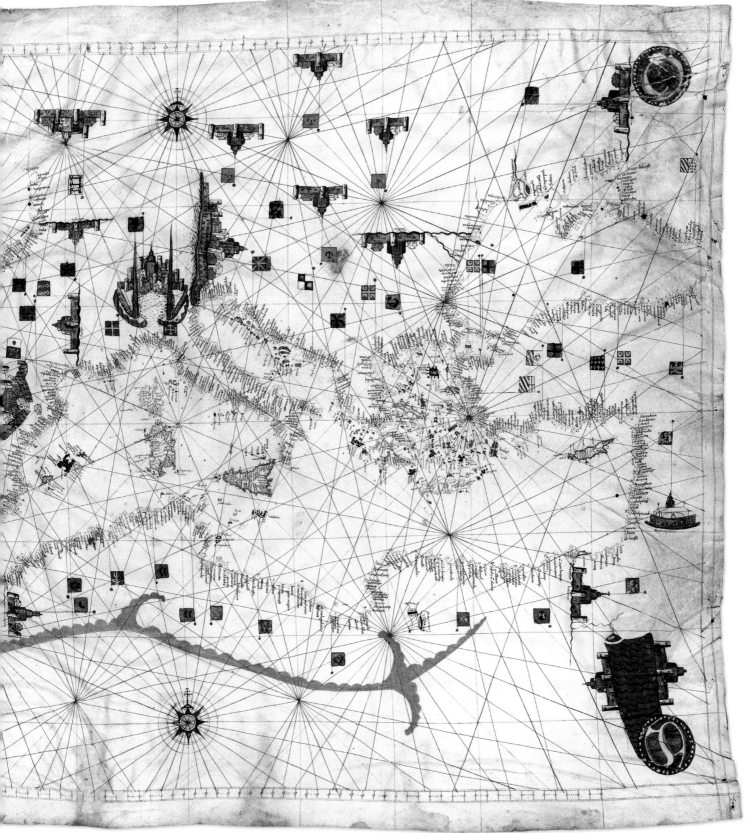

Among the earliest maps to be engraved in England, and featuring the first dated representations of the Spanish Armada of 1588, this map depicts the track of the fleet around Britain. In 1588, King Philip II of Spain sent 130 vessels to invade England. Attacks on the fleet by English naval forces eventually forced the Spanish commander, the Duke of Medina Sidonia, to signal for a return to Spain, but bad weather forced the Armada right around the British Isles. The Spanish suffered very heavy losses: around 20,000 people were killed and 51 ships lost. The defeat of the Spanish Armada – a huge victory for England, which only emerged as a maritime nation towards the end of the sixteenth century – was quickly put to use. English and Dutch propaganda claimed that it was a providential Protestant wind that had maintained Elizabeth I on the English throne. These maps were produced by one of the first English engravers, Augustine Ryther (died 1593), to accompany a textual account by the Florentine mercenary Petruccio Ubaldini, which Ryther himself translated from Italian. At the start of the book, before Ubaldini's text begins, Ryther set a verse echoing the themes of divine will so prominent in the decades after 1588: 'Who list to heare and see what God hath donne/For us, our realme, and Queene against our foe,/ Our foe the Spaniard proud, let him o'rerun/ This little booke, and he the truth shall know:/ The place, the time, the means expressed be/ In booke to read, in graven maps to see.'[5] And, although the account was published two years after the event, Ryther reflected, 'I doubt not but it will breed some pleasure, because the remembrance of pleasures passed are always delightsome.'[6]

*The track of the Armada around Britain and Ireland*, drawn by Robert Adams, engraved by Augustine Ryther, 1590

On 21 October 1805, as he walked on the quarter-deck of the *Victory* during the Battle of Trafalgar, Vice-Admiral Horatio Nelson was fatally wounded by a gunshot and died below deck at 4.30 p.m. The battle was a decisive naval victory for the British and turned out to be the last major fleet action of the Napoleonic Wars (1803–15), but news of Nelson's death overshadowed the success. Sailors wept as word of his demise moved through the fleet. Britain went into national mourning. The funeral, which took place over five days, was the greatest state occasion of the period. By the time of his death, Nelson was a major celebrity, and, in an era in which consumer goods were increasingly available to people of more moderate means, there was money to be made. There was a boom in commemorative artefacts and, along with portraits of Nelson and scenes of his

death, the Trafalgar battle plan became something of a stock image that was used and re-used on paper, metal, textile, ceramics and glass. Through objects like this, Nelson and Trafalgar made their way into domestic spaces, and the battle plan emerged as a symbol of both triumph and loss. This jug was made using a technique known as transfer printing that allowed decorated ceramics to be mass-produced. The plan shows the curved line of the combined French and Spanish fleets and the division of the British fleet – the vanguard led by Nelson in the *Victory* is to the left; the rear-guard led by Vice-Admiral Collingwood to the right. Not all such objects were produced with great care: this jug, for example, incorrectly gives Nelson's age at his death as 48 and the key of letters and numbers set next to the British vessels does not appear to refer to anything.

One element of a commission to mark the centenary of the First World War, this globe, which is also a football, was one of many made to be played with by community organisations across the UK. The choice of a football is important, as it commemorated the informal kickabouts that took place during unofficial ceasefires on the Western Front in December 1914. Part of a larger project, the globe focuses on a moment of cooperation, on the creation of understanding – however fragile and short-lived – and a sense of shared humanity. The artwork as a whole – recordings made of the groups playing with the globe-balls and then shared over social media – is a powerful evocation of these themes. That it is a globe made using satellite imagery is important too: it was intended as an allusion to the iconic 'Earthrise' photograph taken by astronaut Bill Anders from the spacecraft *Apollo* 8 in 1968 that shows the Earth floating in the vastness of space and has become a symbol of the fragility of our planet. What is significant here is that these twin resonances, articulated in the context of increasing insularity and nationalism and climate breakdown, make it clear that commemoration is not so much about the past as it is about today and the uncertainties of the future.

Football, made as part of *One World*,
Mark Wallinger, 2018

D

IS FOR

DISPLAY

While we do not know whether, where and how these particular maps were displayed, we do know that they were designed and made with display in mind. But why display a map? There are a number of possible reasons: to advertise, using maps that seek to attract attention through colour and style; to show off wealth and knowledge, by way of maps designed to impress through their sheer size; to teach, illustrating patterns to help familiarise learners with new information. The purpose of the display is linked to the material nature of the object too, whether made at the greatest expense for a limited audience or produced cheaply as an item that was never intended to last.

Printed on multiple sheets and then pasted onto a linen backing, wall maps like these of the continents Europe, America, Asia and Africa were hung in seventeenth century interiors, unprotected by glass and exposed to dirt, smoke, soot, sunlight, damp and changing temperatures. Such a life makes these maps rare survivals. This set was designed by prominent Dutch cartographer Willem Blaeu. For wealthy Europeans in the seventeenth century, knowledge of global geography resulted from and was celebrated as part of imperial venture. Blaeu's wall maps could be found on the walls of the Dutch East India Company and state offices, as well as in the homes of wealthy merchants. Buyers could choose whether or not to purchase the map surrounded by the text panels featured on these examples, which describe the landscapes, produce and people of different regions. Blaeu's wall maps were even copied, with impressive attention to detail, by Italian publishers who knew that such products would be popular in Italy too, giving merchants and aristocrats opportunities to display their knowledge and wealth.

*Wall Maps of the Continents*,
Willem Blaeu, around 1661

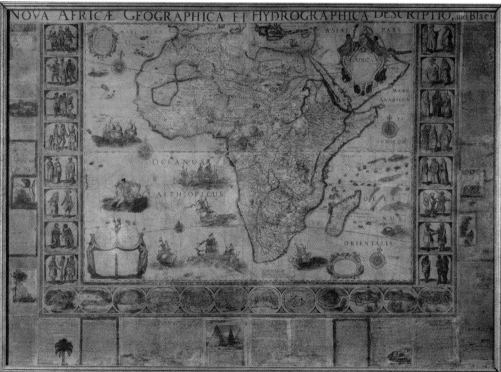

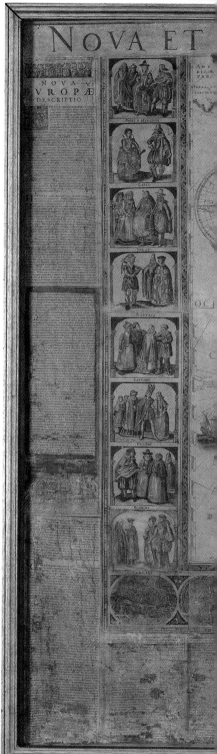

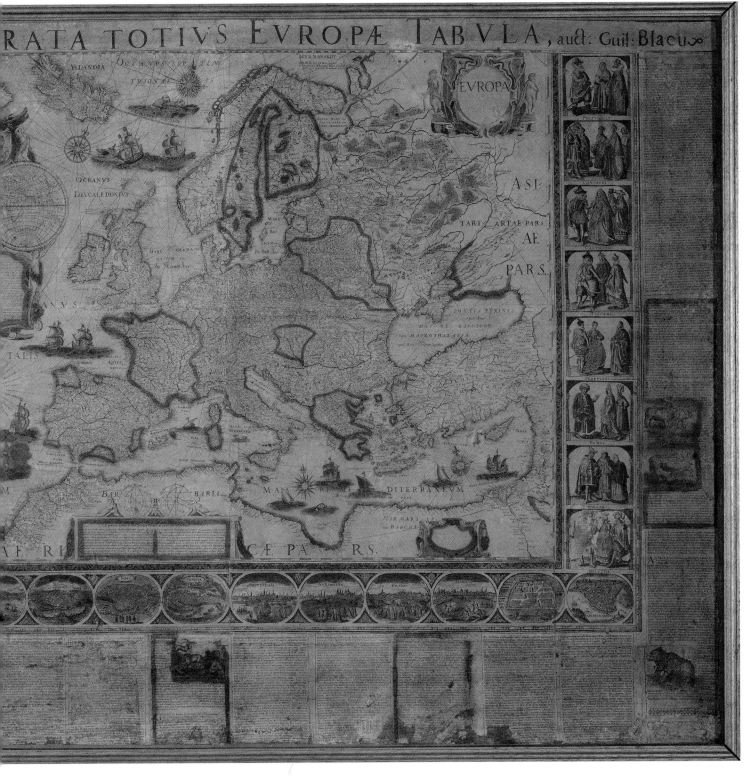

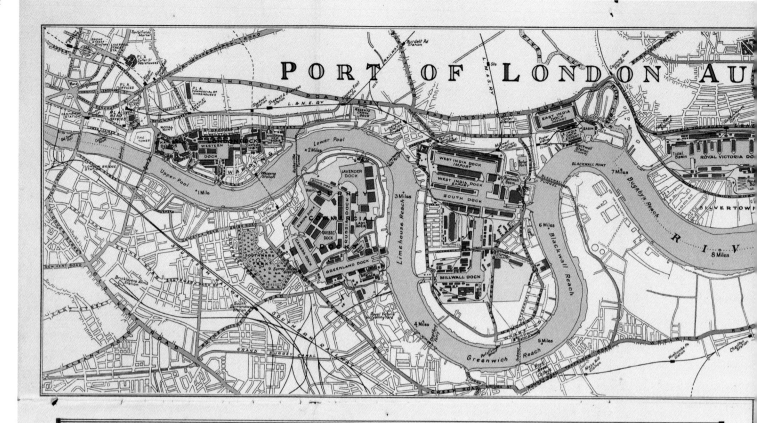

# THE PORT OF LONDON AUTHORITY

IS A PUBLIC TRUST FOR PUBLIC SERVICE AND PROVIDES EVERY

PORT FACILITY FOR LONDON'S VAST IMPERIAL AND FOREIGN TRADE

## OVER ONE THIRD OF THE TOTAL OVERSEAS TRADE OF THE UNITED KINGDOM FLOWS IN AND OUT OF THE PORT OF LONDON

**HEAD OFFICE**

Trinity Square,
E.C.3.

Telephone:—Royal 2000.

**DISTRICT OFFICE**

Theatre Royal Chambers,
New Street,
Birmingham.

Telephone:—Midland 2093.

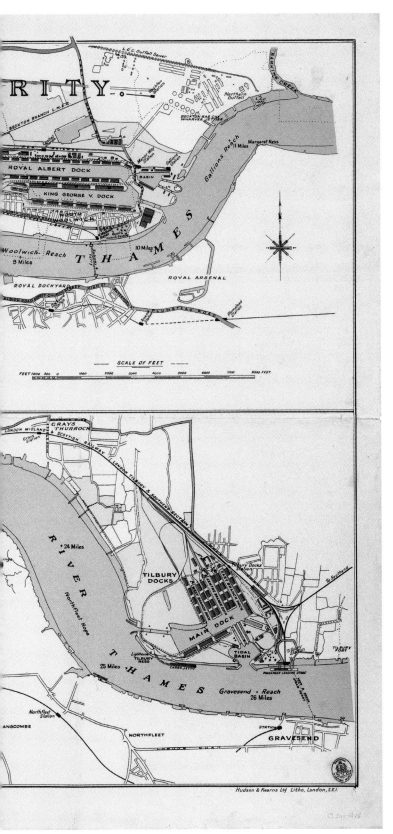

This poster promoted the Port of London and its governing body, the Port of London Authority (PLA), in the decade that would be one of the port's busiest. It illustrates the extent of the dock systems in the River Thames, from St Katharine Docks near the Tower of London, to Tilbury Docks some 25 miles downstream. Having been extensively damaged during the Second World War, the Port of London underwent major reconstruction. Advertising itself was especially important because, alongside reconstruction work, the Authority was developing major plans for improvement and extension, which required huge investment. Encouraging shipping through particular ports, especially following wartime disruption, was a constant preoccupation. This is how ports generate income, after all. The Port of London was also a destination: river and dock cruises, which had started up again in 1948, took tens of thousands of sightseers through it each year, despite major problems with pollution. It was at its busiest in the 1950s but by the late 1960s, just a decade later, the smaller docks were already being closed. The impact of transporting goods in containers, known as containerisation, meant that bigger vessels were more regularly employed and there was less need for the dock systems further up-river.

*The Port of London*, Port of
London Authority, 1950

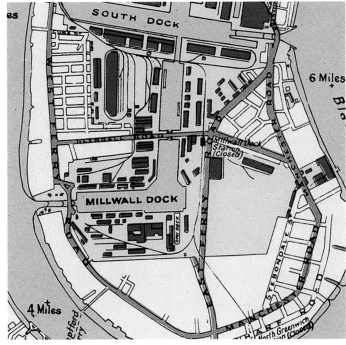

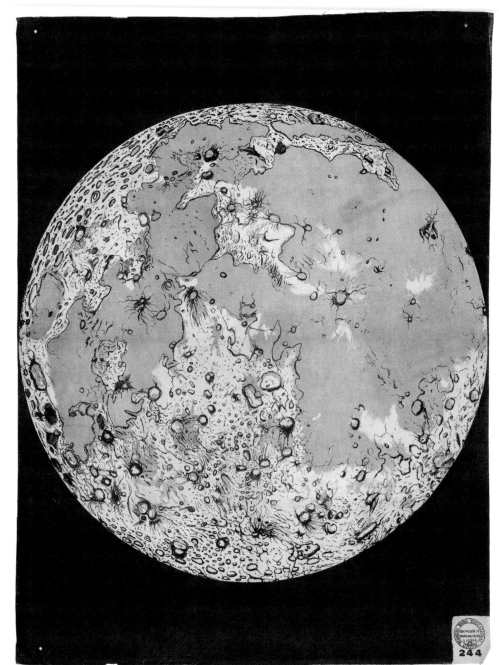

*The Moon, telescopic appearance,*
*3ft diameter*, Working Men's
Educational Union, 1850–60

This map of the Moon, printed on cheap, cotton cloth, was made for display during lectures on astronomy in the mid-nineteenth century. It was produced by the Working Men's Educational Union (WMEU), which was founded in 1853 to provide resources for evening lecturers as a way of supporting mutual instruction – that is, working people teaching each other. At the time, lectures were an important form of entertainment, valued for their educational content but also for elements of show: models, images and even live experiments (topic permitting) were common parts of a lecturer's repertoire.

Suggesting that lecturers could not 'command an audience merely by the allurements of their words', but that the need to illustrate was 'a tax both upon the time and purse of the lecturer', the WMEU produced over 400 different coloured wall hangings that could be purchased individually or in sets.[7] This wall hanging was part of a series concerned with the Solar System and was made in a period when telescopic observations were not only revealing more details of the Moon's surface, but also giving rise to increasing speculation about its formation.

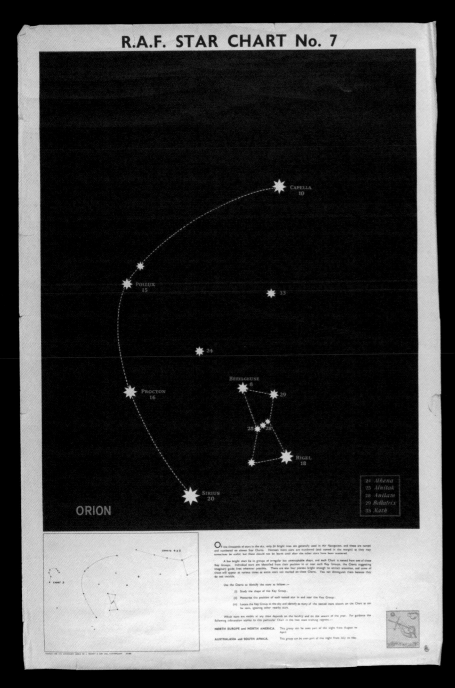

# R.A.F. STAR CHART No. 7

ORION

*R.A.F. star chart No. 7,*
H.M. Stationery Office,
1941

The urgent need for air navigators and pilots at the beginning of the Second World War led to the rapid expansion of training by the Royal Air Force (RAF) for new air crew. Although at the start of the War those learning to navigate did not receive instruction in astro-navigation, it soon became clear that more extensive training was needed. The Navigational Training Branch of the Air Ministry published a series of 11 charts. They were designed for display in places frequented by air crews, so that they could develop familiarity with the principal constellations and the 24 bright stars judged to be most useful for navigators. With stars named and numbered, and with any fainter stars stripped out from the black background, the charts were a visually striking aid. They also offered suggestions for how they could be used to learn the named stars (study, memorise, locate) and how to distinguish planets from stars, as and when they appear ('they do not twinkle'). As *Tee Emm*, the Air Ministry Training Magazine explained, 'it is hoped that their daily study will be of great help to all who have to navigate', since 'you must have a good nodding acquaintance with the stars'.[8]

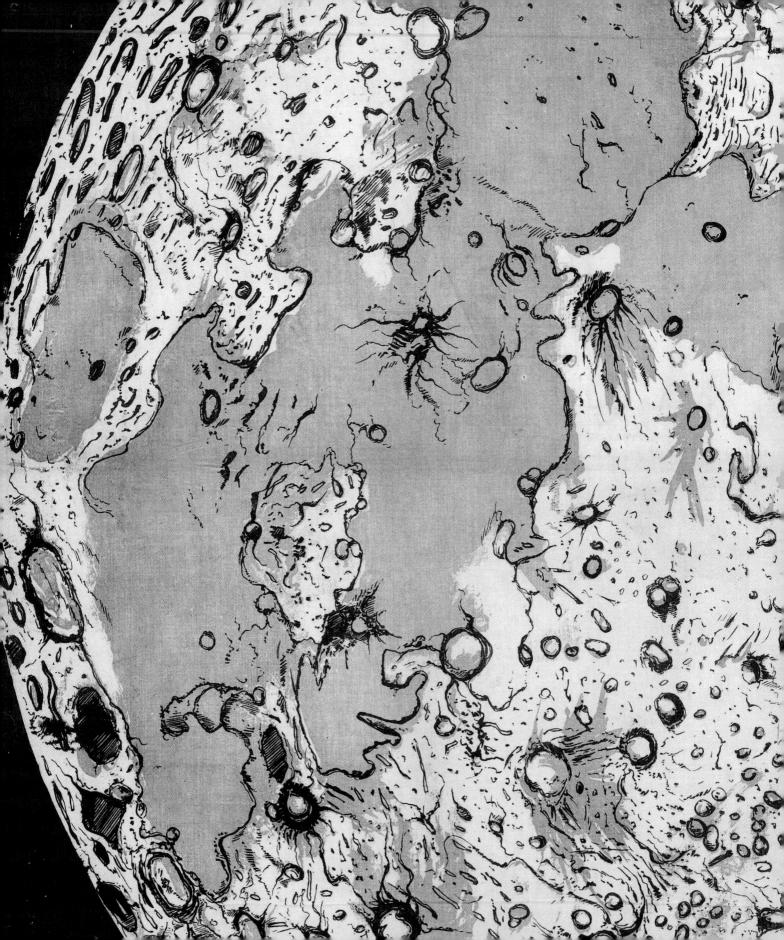

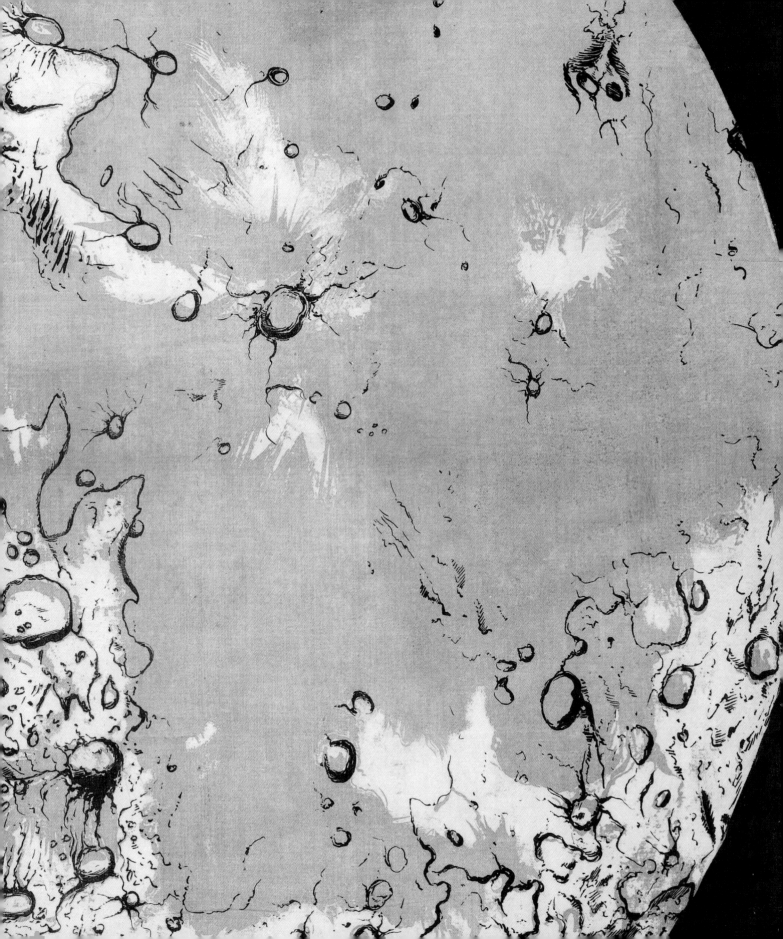

# E IS FOR ENGRAVING

beg, bag, fig, age, big, had, cab, dab. Cooney.

cyz. abcdefghijklmnopqrstuvwxyz. the quick brown fox jumped over the lazy dog. oooooo
ssion, minimum, radio mast, wireless telegraph station, woolloomooloo, wooroorooka, mississippi. s
EFFFFF GGGGG HHHHH JJJJJ KKKK KKKKLLLLLL MMMM MM MMM
QQQQQ RRR RRRRR SSSSSS SSTT TTS
YYYYY ZZZZZ ABC EFGH KLMN PQRST WXYZ UVW CG Q.
AZY DOG. ENGRAVING DEPARTMENT C.B.H ENGLAND WALES SCOTLAND A.
SSIPPI. WOOROOROOKA. SSEX OM ENGLISH ADMIRALTY SURVEYS D.
Channel, Fox Channel, Gulf of Guinea, Hudson Lake, Huron River, Japan Sea, Kennemundy, R.L.
Quarantine Bay, Raasay Sound, St.George Channel, Tor Bay, Ura, Victoria Inlet, Wash, Pangui, R. re
Water Springs. Bearings refer to the True Compass and are given from Seaward (thus:–126° etc.)
93333333333.444444444444444444444.55 5555555555 555555555555.
77777777777.88888888 88888 88 888. 99999999999999999
890.1234567890. 1234567890.1234567890.1234567890. 2222.3333.5555.6666.88888.99999.
0, 1146, 47, 238, 5690, 63, 412, 1952, 75, 936, 1048, 27, 33, 999, 3639, 69, 369, 8957. 26, 963, 4393, 32, 693. 11-1-52.
235 235 235 235. 22222222222 3.333333333333333333333222222222 25555555555
5455 65 66 75 72 73 74 75 81 82 83 84 85 9 92 93 94 95 101 102 103 104 105 22 23 82 25 82 23 24 83 23 55 83 32 15
41 41 42 43 51 51 53 64 61 62 71 71 73 84 81 83 91 92 93 101 101 102 23 85 91 93 101 81 32 23 81 35 21 83.21.
oooooo 111111 1111111 1.iiiii iii 11 iii iiii ffffff ffff ffffffffffff
rr. ttttttttt tttt tttttt.uuuuuuu uuu uuuuu.nnnnnnnn nnnnnnn.
n.vvvvvvvv vvvvv vv.w ww wwwwwww www ww.xxxxxxx xxxxxxxx.
zkkkkkkkk kkkkkkkkk.dd d d d ddddddd.b bb bbbbb bbb bbbbb.
ccccccc ccccccc c.eeeee e ee ee e eeeee.oooooo ooooooooo.
ggg gg gg g gg gg g g.aaa aa asssssss ss.ggg gg g g.stag. stag. stag.
q Rr Ss Tt Uu Vv Ww Xx Yy Zz. abcdefghijklmnopqrstuvwxyz.abcegjkmnopqsuvwxyzd.
ark Denmark, Egypt Egypt, Florida Florida, Greece Greece, Hungary Hungary.
Malta Malta, Norway Norway, Orkney Orkney, Portugal Portugal, Queensland Queensland
raine Ukraine, Venezuela Venezuela, West Indies West Indies, Yugoslavia Yugoslavia.
NDINGS IN FATHOMS, SOUNDINGS IN FEET, CONSPICUOUS TO THE NAVIGATOR. NOTES.
OBJECTS, TRUE MERIDIAN. (Under Eleven in Fathoms and Feet). Fathoms and Feet. Eleven
og Siron, (Under Eleven in Fathoms and Feet). SOUNDINGS IN FATHOMS

iiiiiiiiiiiiiiiiiiiiiiiiiiiiii till tilt lit little
fffffffffffffffffff ff ff nnnnnnnn
m mm mm m mm m m m m mm rrrrrrrrrrr
ooooooo oooo o o ooo ppppppppp
bb bbbb bbbbbbbbb aaaaaaa
cecececececececececec cccccccc cccccccc
reduced receded accident defeated accede good

Occ.W.G. 3 sec.
140f¹ 13.10M.

3170

Occ.W.G. 3 sec.
140f¹ 13.10M.

RAILWAY FROM PORT ELIZABETH

RAILWAY FROM KOROGEN

Cables 10
(6047 ft)       5       0

Feet 1000     0       5000

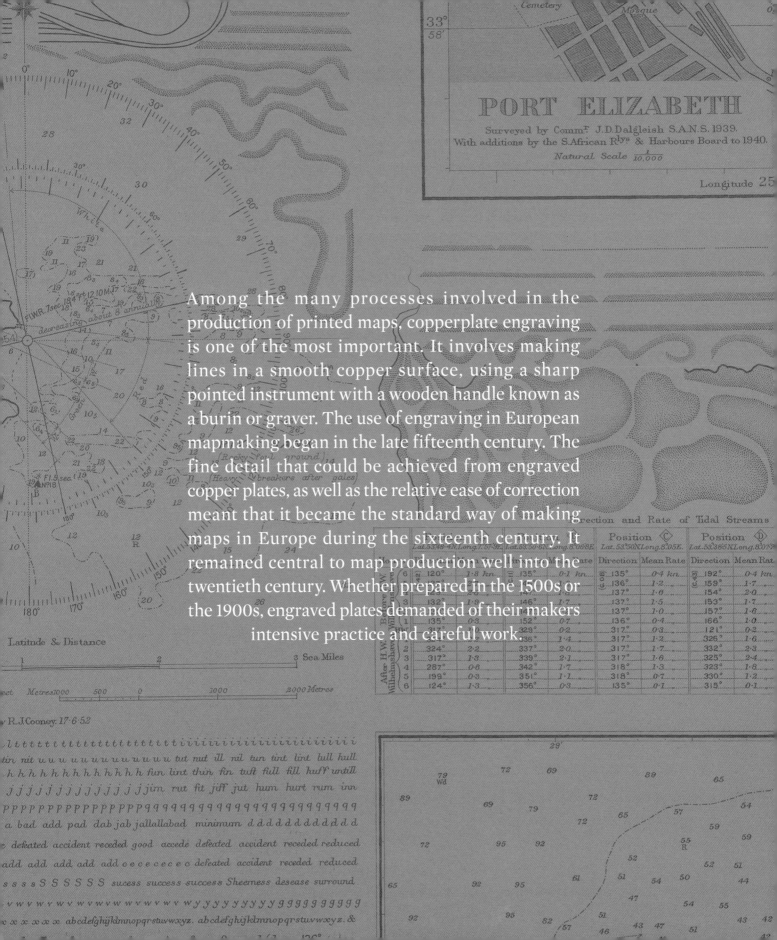

Among the many processes involved in the production of printed maps, copperplate engraving is one of the most important. It involves making lines in a smooth copper surface, using a sharp pointed instrument with a wooden handle known as a burin or graver. The use of engraving in European mapmaking began in the late fifteenth century. The fine detail that could be achieved from engraved copper plates, as well as the relative ease of correction meant that it became the standard way of making maps in Europe during the sixteenth century. It remained central to map production well into the twentieth century. Whether prepared in the 1500s or the 1900s, engraved plates demanded of their makers intensive practice and careful work.

PORT ELIZABETH

Surveyed by Comm$^r$ J.D.Dalgleish S.A.N.S. 1939.
With additions by the S.African R$^{lys}$ & Harbours Board to 1940.

Natural Scale $\frac{1}{10,000}$

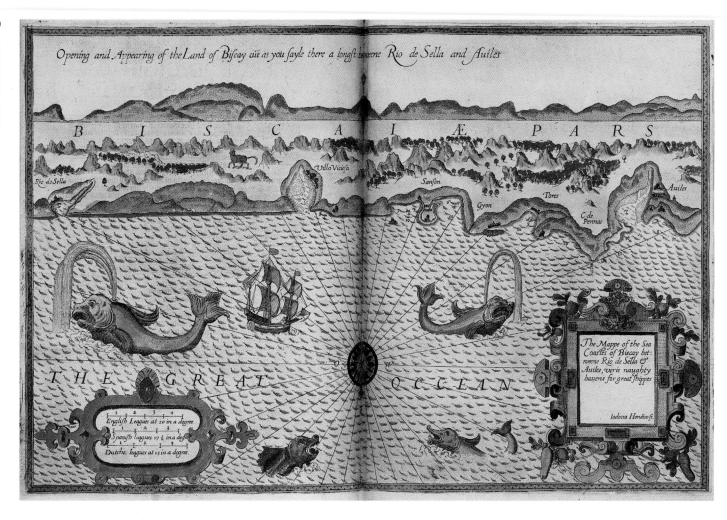

This chart was engraved in London by Jodocus Hondius (1563–1612). Hondius arrived in the city as a refugee, having fled religious conflict in his native Ghent, then in the southern Netherlands. Already skilled in the art, he was able to find work engraving cartographic plates, during a period in which there were very few English practitioners. It was a highly skilled activity. It took years of practice to achieve the manual control necessary to make lines of consistent depth and thickness, smooth curves and regular waves, contours and shading. Among other things, Hondius worked on the prestigious English translation of Lucas Jansz Waghenaer's sea atlas, *Spieghel der Zeevart*, which was published as *The Mariners*

*Mirrour* in 1588. The atlas was so influential that bound volumes of charts and sailing directions became known among English navigators as 'waggoners', a nod to Waghenaer. Hondius, who moved to Amsterdam in 1593, went on to develop a reputation as one of the most important cartographic engravers of the early seventeenth century.

'The Mappe of the Sea Coastes of Biscay', *The Mariners Mirrour*, engraved by Jodocus Hondius, 1588

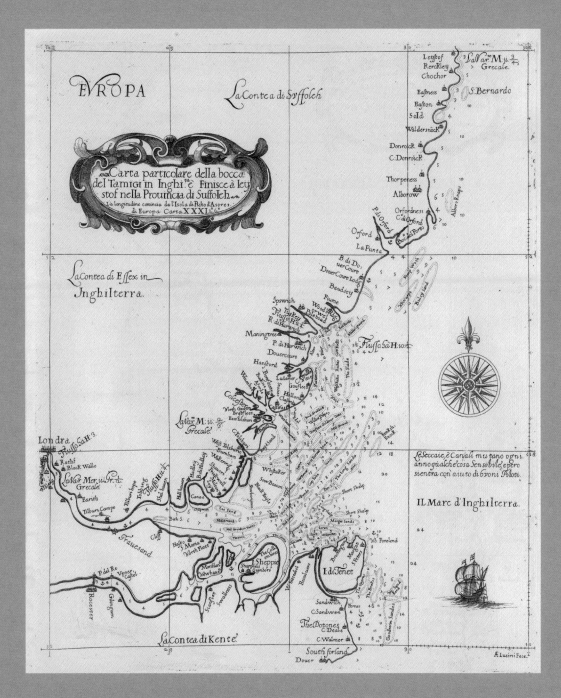

Antonio Francesco Lucini engraved this map for Robert Dudley's second edition of *Dell'Arcano del Mare* ('Of the Mystery of the Sea'). Its first edition, published in 1646, was said to be the first sea atlas compiled by an Englishman. Dudley completed the work while in exile in Florence after his claim to be the legitimate heir of his father, the Earl of Leicester, was rejected. Maps engraved and published in Rome, Florence and Venice were very highly regarded in Europe during this period and it was in Tuscany that Dudley's volumes were made. Each of the two unwieldy volumes weighed around 7.5 kilograms. It was published by Lucini in 1661 after Dudley's death. Lucini noted in his preface to the new edition that the engraving took 12 years to complete and required the use of 5,000 libbra (Florentine pounds – around 1,600 kilograms) of copper. The style of the finely engraved charts in *Dell'Arcano del Mare* is widely understood to be a result of Lucini's work interpreting Dudley's maps onto copper. Indeed, engravers were artists in their own right, though in later centuries they would have to fight for this recognition.

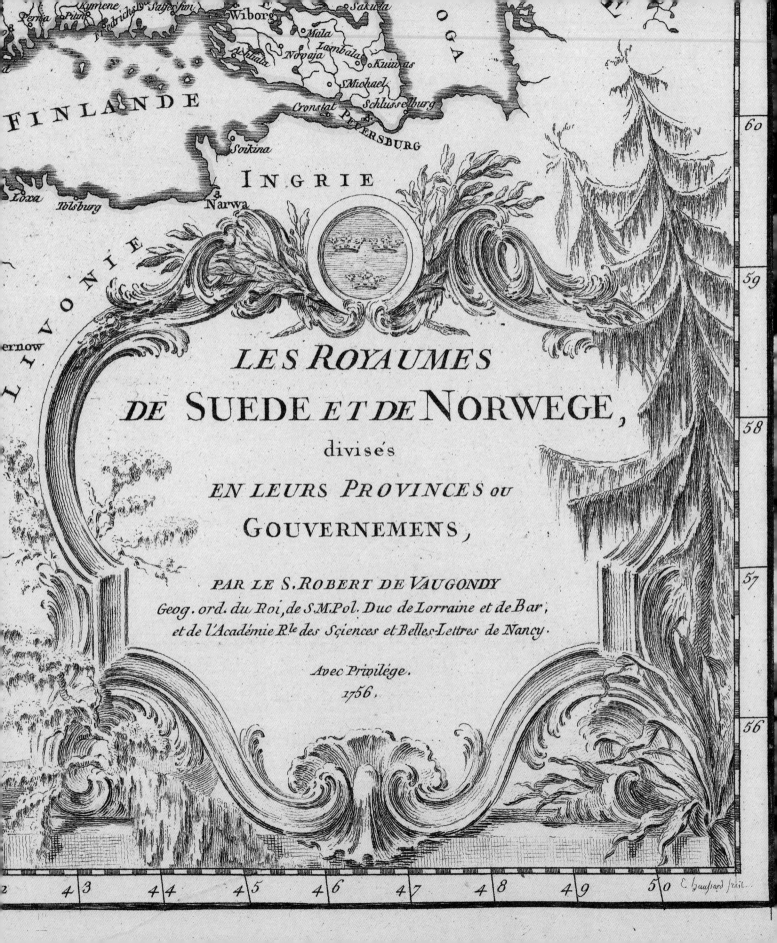

FINLANDE

Kymene · Pitua · Fredrichs · Saljerfion · Wiborg · Sakuta
Mala
Novaja · Lambala
Kuimas
S.Michael
Cronstat · PETERSBURG · Schlusselburg
Soikina

LOGA

INGRIE

Narwa

LIVONIE

Loxa · Tolsburg

ernow

## LES ROYAUMES
### DE SUEDE ET DE NORWEGE,
*divisés*

## EN LEURS PROVINCES ou
## GOUVERNEMENS,

PAR LE S. ROBERT DE VAUGONDY
Geog. ord. du Roi, de S.M.Pol. Duc de Lorraine et de Bar,
et de l'Académie R.le des Sçiences et Belles-Lettres de Nancy.

Avec Privilége.
1756.

60

59

58

57

56

42 · 43 · 44 · 45 · 46 · 47 · 48 · 49 · 50

E. Haussard fecit.

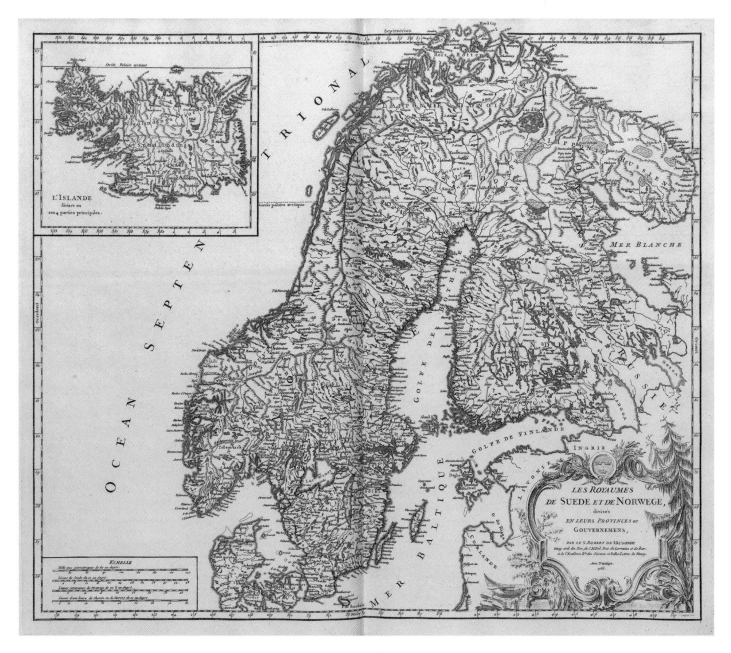

Just visible in the bottom right corner of this map (see detail) are the words 'E Haussard fecit', or 'made by E Haussard'. This reveals that the cartouche, or decorative title piece, of this map from Gilles Robert de Vaugondy's *Atlas Universel* (1757) was engraved by Elisabeth Haussard (1700–1804). Haussard worked alongside her sister Marie Catherine as an engraver in eighteenth-century France, having been taught the art by her father. Both women worked on a number of major technical and scientific publications. As often happened in larger map-publishing projects, different engravers worked on different parts of the *Atlas Universel* and even on different parts of the same map. Elisabeth Haussard specialised in decorative cartouches using imagery associated with particular regions, like the tall pine tree on this map of the Scandinavian kingdoms. Haussard's work, recognisable thanks to her faintly engraved name, is one of the more visible examples of women's skilled involvement in map-making from this period, at a time when many women had active roles in the book and print trades.

'Les Royaumes de Suede et de Norwege…', *Atlas Universel*, cartouche engraved by Elisabeth Haussard, 1757

This is a print from a 'practice plate' by a 16-year-old apprentice engraver, Roy Cooney. It was made in the early 1950s at the Hydrographic Department (now the United Kingdom Hydrographic Office) in Taunton, Devon, one of the most intensive chart-producing organisations in the world. It shows how particular skills were learned early in the apprenticeship. After making their tools and learning to draw letters back to front (since printing gives a reverse image of what is on the plate), apprentices started with short, straight, even lines in the first weeks, before progressing to curved lines, different styles of lettering and details of increasing complexity. Finally, these elements were combined into the small sample chart at the end of the sheet. It was only after completing a practice plate that an apprentice would be allowed to make small corrections to the plates from which navigational charts were published. After a six-year apprenticeship, they would be permitted to engrave new charts as a journeyman engraver. The Hydrographic Office had worked with engraved copper plates since the publication of its first chart in 1800. For many decades the Hydrographic Office used newer, cheaper methods of chart production alongside engraving, but the engraving department was finally disbanded in 1981 out of preference for these more modern technologies.

Practice plate and engraving
tools, Roy Cooney, 1951–52;
date unknown (tools)

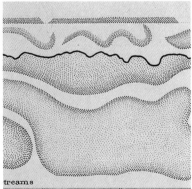

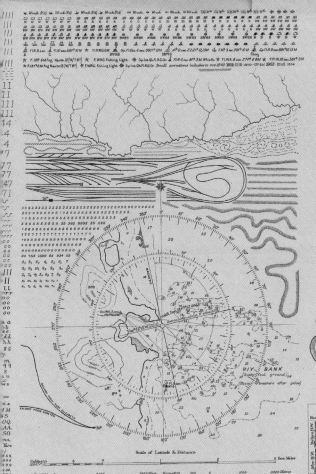

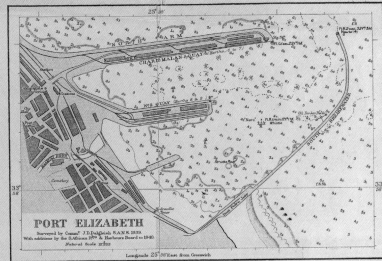

## PORT ELIZABETH

Surveyed by Comm.ʳ J.D.Dalgleish S.A.N.S. 1939.
With additions by the S.African Rly.s & Harbours Board to 1940.
Natural Scale 1:12,355

Longitude 25° 38' East from Greenwich

*Engraved by R.J.Cooney, 29·8·52*

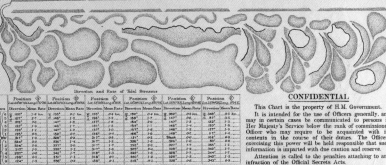

### CONFIDENTIAL

This Chart is the property of H.M. Government.

It is intended for the use of Officers generally, and may in certain cases be communicated to persons in Her Majesty's Service below the rank of commissioned Officer who may require to be acquainted with its contents in the course of their duties. The Officers exercising this power will be held responsible that such information is imparted with due caution and reserve.

Attention is called to the penalties attaching to any infraction of the Official Secrets Acts.

*The quick brown fox jumps over the lazy dog.*

*The quick brown fox jumps over the lazy dog. Bearings refer to the True Compass and are given from Seaward (thus—126° etc.) so All other heights are expressed in feet above High Water Springs. For Abbreviations see Admiralty Chart 5011. Small Corrections.*

*with lift hymn limit kilt sulad*

*body door hood joint would iron void*

*success Sheerness bliss kriss*

*London. Published at the Admiralty 1st May 1952, under the Superintendence of Rear-Admiral A.Day, CB,CBE,D.S.O., Hydrographer.*
*Prepared by the Hydrographic Dep.t of the Admiralty 1st Feb.1951, under the Superintendence of Rear-Admiral A.Day, D.S.O., Hydrographer.*
*Issued for Fleet Purposes by the Hydrographic Dep.t of the Admiralty 1st Jan.1951 under the Superintendence of Rear-Admiral A.Day, CB, D.S.O.*

COMPILED FROM ADMIRALTY SURVEYS TO 1952. FROM A UNITED STATES GOVERNMENT CHART TO 1954
SURVEYED BY OFFICERS AND MEN OF H.M.S."SEAGULL" 1896. FOR ABBREVIATIONS SEE CHART 10
BRITISH ISLES    THE MEDITERRANEAN SEA WESTON-SUPER-MERE S S S S
THAMES ESTUARY AND APPROACHES TO THE NORTH SEA

## THE EASTERN INDIAN OCEAN OCSDOCSDN

### APPROACHES TO CARTWRIGHT HARBOUR

### UNION OF SOCIALIST SOVIET REPUBLICS

(1). German Admiralty charts are constructed on the Mercator projection. Since 1924, the mid-parallel of each chart has been used in computing the natural scale, but on older charts the natural scale refers to the lowest parallel. For the coastal charts of German waters the parallel of Lat. 53° 05' N. has been adopted. Direction Finding and Great Circle charts are constructed on the Gnomonic projection.
(2). On many German smaller scale charts, a small inset is shown giving the larger scales within the area of the chart. These larger scales include British and Foreign, as well as German charts
(3). Most German charts are engraved on copper plates and are printed in black only. A certain number, mostly large scale harbour sheets are printed in colours, a grey-green being used for the land, and a series of graded blue tints for the shallower water.

*First Practice Plate Engraved by R.J.Cooney. Commenced Oct.1951. Completed Dec.1952.*

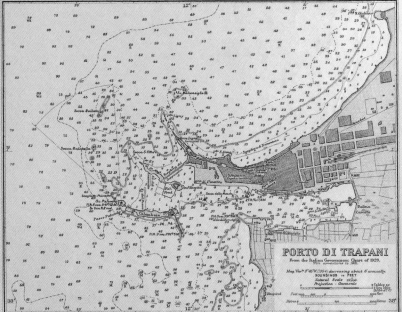

## PORTO DI TRAPANI

From the Italian Government Chart 1929.
With corrections to 1951.
Mag.Var.ⁿ 3°45'W.(1954) decreasing about 6' annually.
SOUNDINGS in FEET
Natural Scale 1:22,500
Projection – Gnomonic

*Engraved by R.J.Cooney. 22·12·52*

G.298·1/5

Fake, forgery, facsimile. These are the three 'Fs' of maps that are not quite what they seem: forgeries, sometimes called 'deliberate fakes', are made with an intent to deceive; fakes are a broader category – objects interpreted to be original, regardless of the intent of the maker; facsimiles, meanwhile, are produced as exact copies for historical or scientific purposes and generally not intended to pass, or be received as, originals. Some of the objects here were previously thought to be genuine but have since been identified as fakes. Research into counterfeits and imitations often reveals them to be remarkable in their own right. In many cases, considerable care goes into producing a fake, either by making use of traditional techniques and tools to create something convincing or by using new technologies of copying. Other objects, facsimiles rather than original maps and globes, were never intended to deceive and have their own stories and purpose.

Goblet globes, highly valued in early modern Europe, have always been collectors' items. This globe is a reproduction, made using a technique known as electrotyping, an electrochemical process developed in the nineteenth century. It involves the even transfer of particles of metal to a mould by an electric current, enabling extremely precise replicas to be made. It was, for a time, extremely popular with museums. The Victoria and Albert Museum in London promoted a programme for reproducing important works of art 'for the benefit of museums of all countries', with the idea that replicas could be shared with international, regional and local museums to facilitate study. Electrotyping companies, keen to advertise their prowess as makers of incredibly precise copies, tended to stamp the objects they produced for study so there was no way they could pass as the real thing. This globe, not stamped in this way, is thought to be a 'deliberate fake', intended to be sold and bought as an early seventeenth-century original rather than as a nineteenth-century copy made using the latest technology.

**Celestial goblet globe,**
**unknown maker, 19th century**

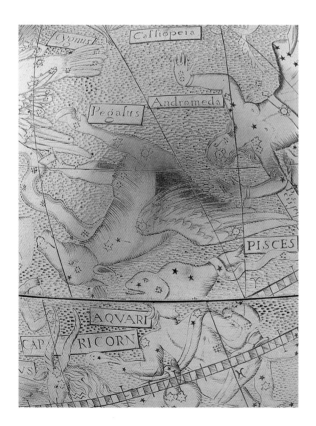

This is a revolving clockwork globe. It is driven by a watch mechanism that sits inside the sphere. Wound with a key at the North Pole, it would have run for 30 hours and the time was indicated by a small pointer in the shape of the Sun. The gores, or paper segments pasted over the sphere, were published in 1620 by the prominent seventeenth-century Dutch cartographer Johannes Janssonius and the watch mechanism was made by German clockmaker Johann Tomas Seyler around 1650. There is evidence of wear, which suggests that the mechanism was inside a watch for a considerable time before being adapted to make the globe turn. At first glance, the globe, like many ornamental globes, appears to be carried by a figure representing the Titan Atlas from Greek mythology. It has been suggested, though, that the form appears to have iconographic elements in common with Christ crucified and may have been removed from a crucifix and repurposed as a globe stand.[9] The work was carefully done, however; traditional tools and techniques were used to adapt the mechanism and both it and the globe gores are roughly contemporary. Some fakes, then, with original parts sourced separately and subsequently brought together by a more recent artist, are Frankensteins.

Terrestrial clockwork globe, gores by Johannes Janssonius, 1620, watch mechanism by Johann Tomas Seyler, around 1650, assembled by unknown maker, around 1930

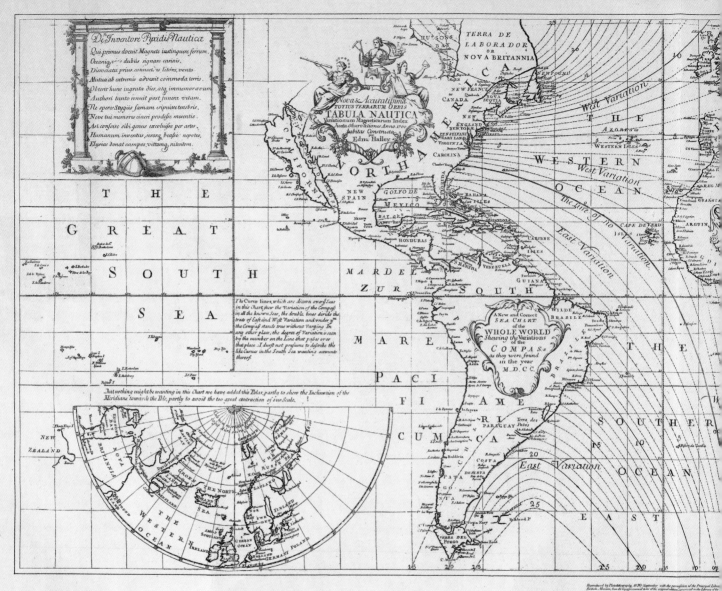

In the second half of the nineteenth century, the ability to make and print photographic copies of line drawings was a new and exciting development, bringing together two scientific innovations from earlier in the century: photography (making images using light-sensitive materials) and lithography (printing using blocks of limestone specially prepared using grease, gum, water and ink). 'The power of taking copies of objects by photography on stone, then etching and printing direct from these copies, is wonderful', enthused the *Scientific American*, while simultaneously worrying about the potential for producing fake banknotes.[10] But photolithography was perfect for reproducing maps and the technique was enthusiastically adopted by the Ordnance Survey. So when George Biddell Airy, the Astronomer Royal working at the Royal Observatory in Greenwich, finally tracked down a map he had been trying to trace in all the various scientific academies in Europe, it was through photolithography that he had it reproduced. The map itself dated from the early 1700s and showed magnetic variation across the world by the second Astronomer Royal,

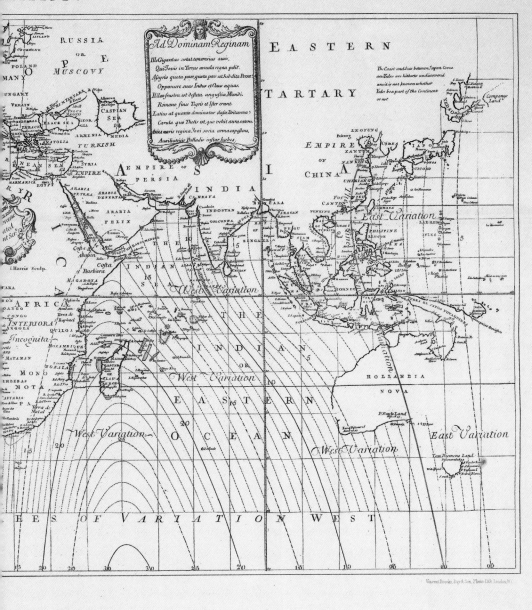

Edmond Halley. It drew together both Halley's own magnetic observations made in the Atlantic at the end of the 1690s and those he had compiled from the logbooks of ships in the Indian Ocean around 1700. Halley pioneered the use of isogonic lines – lines linking points of equal magnetic variation (how far from true north a compass points) – and this depiction of magnetic data is what had made Airy so excited. Evidently pleased with the ability to copy, Airy signed this map, adding his name to past studies of the Earth's magnetism.

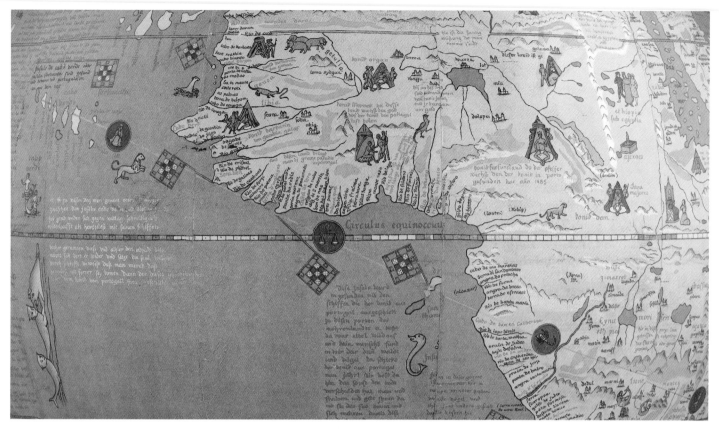

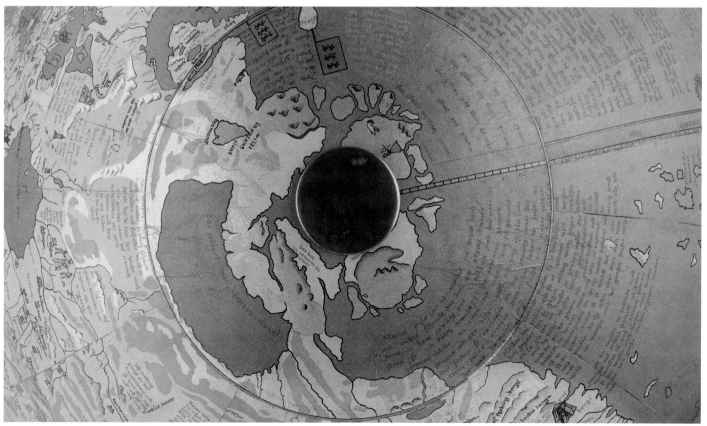

The earliest known terrestrial globe, Martin Behaim's 'Erdapfel', or earth-apple, has long held fascination for geographers and historians. Made in 1492, before Europeans knew of the existence of the Americas, it brings together geographical knowledge from ancient texts and medieval travel accounts. Extensively annotated by Behaim, it gives good insight into a German scholarly world view at the end of the fifteenth century. The first known copy of the globe was made in 1770; copies are still being made today. This facsimile uses gores published in 1908, which were the main way that scholars accessed Behaim's globe for much of the twentieth century. Ernst Ravenstein (1834–1913), who made the gores, gave a close account of how he did it: tracing and photographing an existing facsimile; laboriously comparing those images with the original. He made critical editorial decisions, too: the colours are vibrant, not faded; signs of wear are not present; place names added to the globe after 1492 were omitted; names Ravenstein felt probably were on the globe originally were included. As Ravenstein himself wrote, 'I am fully aware that if the definition of etymologists of a facsimile as an "exact copy or likeness" be insisted upon, the copy or version now produced cannot claim that description.' The gores were produced from the closest study of the globe up to that point. Although only 510 of these copies were produced, their greater accessibility meant that they were the main object of study for those interested in Behaim's earth-apple, rather than the original. In producing his facsimile, Ravenstein departed from the object in an attempt to get back to something that no longer existed and possibly never had. As a critical edition, albeit a spherical one, this globe raises questions about the very meaning of originality.[11]

Facsimile terrestrial globe, gores
by Ernst Ravenstein, about 1920
(original 1492)

G IS FOR GLOBE

People have made globes to represent both the Earth and heavens for thousands of years. Written evidence from around 18 CE describes a terrestrial globe (of Earth), displayed in Pergamon some 170 years previously. The earliest extant celestial globe (of the stars around Earth) is part of a Roman statue dating to around 150 CE and thought to have been based on an earlier Greek original. Traditions of celestial globe-making developed, particularly in the Islamic world, during the medieval period. In Europe around 1500, globes – celestial and terrestrial – started to be produced more widely, typically from cast, engraved metal or from papier mâché and plaster with paper pasted over the top. Over the centuries makers have also experimented with different methods of construction and materials: beaten strips of metal could be fashioned into a sphere; tissue paper or leather were inflatable; glass could be used to model the stars around Earth.

This celestial globe is not signed by a maker but is part of a long tradition of globe-making in the Islamic world that brought together Persian, Indian, European and African astronomical knowledge. Celestial globes were made across the Islamic world, from Spain to India. The earliest that survives has been dated to between 1080 and 1085. This example is constructed from cast metal and made using a complex, multi-stage process of mould-making known as the 'lost wax' method. It demonstrates the excellence of its crafter by allowing the creation of a seamless globe, manufactured as a whole, hollow sphere, rather than as two joined hemispheres. The process was also significantly more time-consuming than casting the object in two halves and is part of what enables us to locate the region in which this globe was produced. Islamic celestial globes cast as whole spheres are particular to workshops in what is now Pakistan and the north-west of India. What is more, the style in which the constellations – 1,018 stars inlaid in silver – have been engraved on the globe are similar to other pieces known to be from the Lahore workshop of Qāʾim Muḥammad and then of his son Ḍiyā al-Dīn Muḥammad, prominent makers of astrolabes and celestial globes in the seventeenth century.[12] Where explicit maker's marks are absent, understanding more about how objects were made can provide invaluable clues to their history.

Celestial globe, unknown maker,
first half of the 17th century

Made from strips of beaten copper formed into a sphere, this globe proposes a new way of looking at the stars. Several astronomers in seventeenth-century Europe suggested alternative schemes for arranging the figures of the constellations, which had been based predominantly on classical mythology for thousands of years in European and Islamic astronomy. Out of concern about the use of pagan imagery, scholars developed schemes for both religious and political alternatives to these figures. Erhard Weigel (1625–99), a mathematician and astronomer from Jena, Germany, developed a new set of constellations in 1661 based on European heraldry: *Ursa Major*, the Great Bear, became an elephant (opposite, top), associated with Denmark; *Aries*, whose symbol is a ram, became the Lamb of God with a pennant, representing the Church; *Pegasus*, the winged horse, became the white horse in the heraldic device of Braunschweig-Lüneburg. Painted flush with the sphere are the more familiar constellations. The new constellations are then shown in relief, standing out from the beaten-metal surface of the globe. Weigel's arrangements of stars did not catch on and neither did this method of constructing globes. But the construction suited Weigel's purposes, allowing the newly proposed and traditional constellations to be shown in relation to each other.

Celestial globe,
Erhard Weigel, 1699

Erhardi
WEIGELII
Cons. Cæs. et Pal:
Prof. Honorar.
Globus Coelestis
corr. et perpe
tuus
Jenæ
1699

At 108 cm in diameter, this was the largest commercially available globe in the seventeenth century. The gores, the paper segments that cover the globe, were made by the Venetian friar, cartographer and geographer Vincenzo Coronelli (1650–1718) in 1688. Coronelli produced globes of various sizes and his work was highly fashionable in Italian, French and German-speaking lands, although his publishing enterprises foundered in the early eighteenth century. Like most European globes, this one is constructed of papier mâché and plaster, with the gores pasted over the top. This particular globe, however, was constructed long after Coronelli had died. The engraved (vertical) meridian ring and a scrap of paper found inside the globe during extensive conservation work in the 1930s, reveal that the gores were mounted in Vienna between 1752 and 1754. The work was led by another Franciscan friar and mathematician, Tobias Eder, who was around 80 at the time, assisted by an engraver, Matthias Heinen. We do not know when the gores came to the Viennese friary, whether they sat in a library for years before someone undertook to mount them or were purchased specifically to be pasted onto an enormous sphere. Regardless of what happened, the globe is an example of the complexity of globe-making and the movement of paper products around early modern Europe.

Terrestrial floor globe, gores
by Vincenzo Coronelli, 1688,
mounted by Tobias Eder and
Matthias Heinen, 1752–54

This tissue-paper inflatable globe was made in Bristol in the nineteenth century. Its box describes how 'it is so portable, it may be carried in the pocket and in one minute be expanded to a circumference of 12 feet'. School teacher and inventor George Pocock (1774–1843) also produced his inflatable globes in silk – one was presented to Prince George, the Duke of Cumberland, as a birthday gift from his parents. An article about Pocock's globes in the *Liverpool Mercury* enthuses of one of them that 'by its superior size, its independency of machinery, its buoyancy of movement, it conveys, without any illustration, a natural idea of the floating orb which it is intended to represent' and extols its use both in the schoolroom and at home, as the basis for polite conversation. Rather optimistically, a pamphlet accompanying the globe suggested inflating it by waving it back and forth. More realistically, Pocock also advertised pumps. Pocock's inventions were many and most involved paper or fabric structures: tents that would serve as portable churches for Methodist preachers without access to large buildings; kites (what he was most famous for at the time in his hometown of Bristol) that would lift people off the ground or pull a special carriage at speeds of up to 25 miles an hour. His globes, the design of which was apparently inspired by the construction of hot-air balloons, brought together his fascination with flight and the possibilities of making three-dimensional objects from paper and fabric.[13]

Inflatable globe, George Pocock, 1830

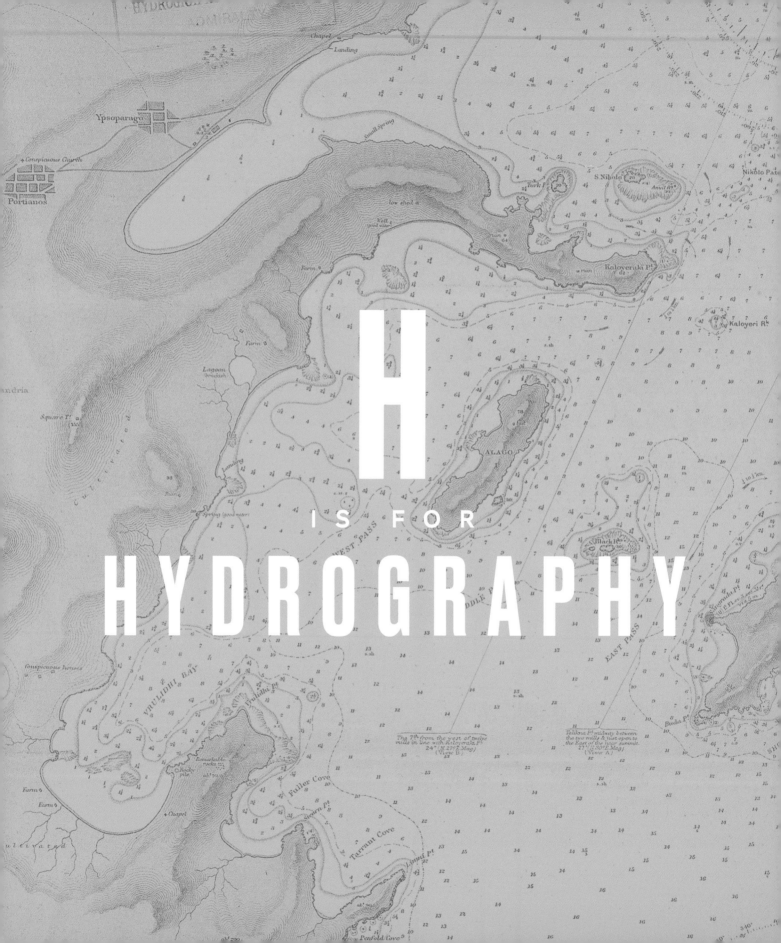

H

IS FOR

HYDROGRAPHY

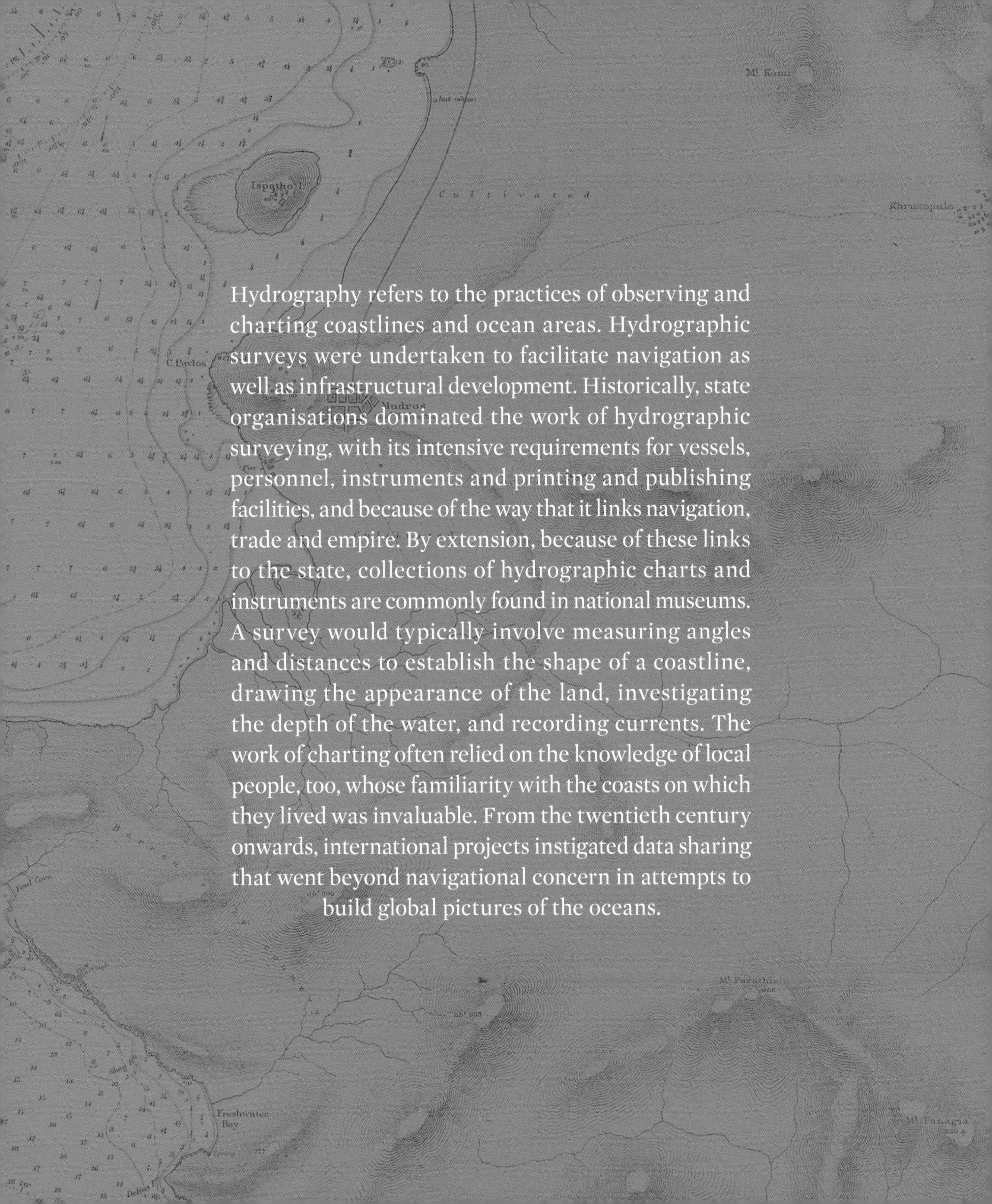

Hydrography refers to the practices of observing and charting coastlines and ocean areas. Hydrographic surveys were undertaken to facilitate navigation as well as infrastructural development. Historically, state organisations dominated the work of hydrographic surveying, with its intensive requirements for vessels, personnel, instruments and printing and publishing facilities, and because of the way that it links navigation, trade and empire. By extension, because of these links to the state, collections of hydrographic charts and instruments are commonly found in national museums. A survey would typically involve measuring angles and distances to establish the shape of a coastline, drawing the appearance of the land, investigating the depth of the water, and recording currents. The work of charting often relied on the knowledge of local people, too, whose familiarity with the coasts on which they lived was invaluable. From the twentieth century onwards, international projects instigated data sharing that went beyond navigational concern in attempts to build global pictures of the oceans.

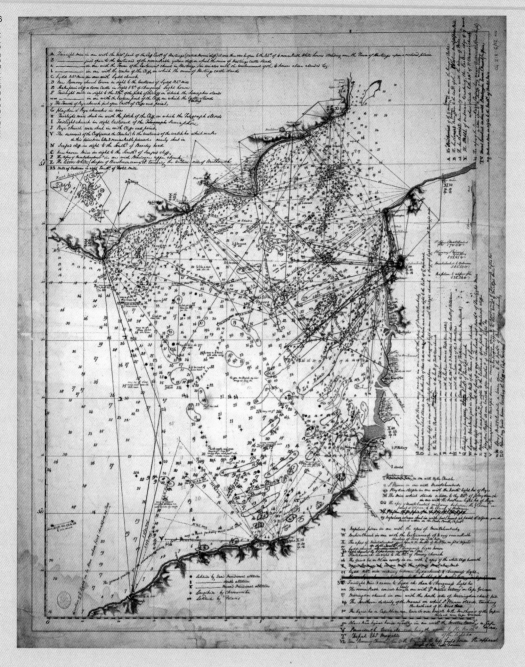

This manuscript survey brings together triangulated measurements between points on land, bearings to prominent coastal points from out at sea, depth measurements, descriptions of the seabed and measurements of latitude and longitude. As such, it shows something of the intensity of work that went into producing hydrographic charts in the earlier nineteenth century. Around the edge is a key to particular lines of sight, key position-finding aids for those at sea – at 'qp' you would see 'Playden steeple in one with the northern lighthouse of Rye'; at 'm', 'the fir trees at Paddlesworth in one with Folkstone church'. Yet describing the navigational difficulties of the English Channel, White emphasised that the often poor visibility meant that the only guide was, in fact, the bottom of the sea.[14] A lead and line, with tallow stuck to the bottom of the lead, allowed surveyors and navigators alike to learn both the depth of the water and what was on the seafloor – sand or mud, shells or gravel. Navigators were in turn able to determine their course using similar techniques. Such seabed investigations, utilising a navigational technique used for centuries, also fed into what would become the science of oceanography.

*Survey of the English Channel east of Beachy Head*, Martin White, 1823

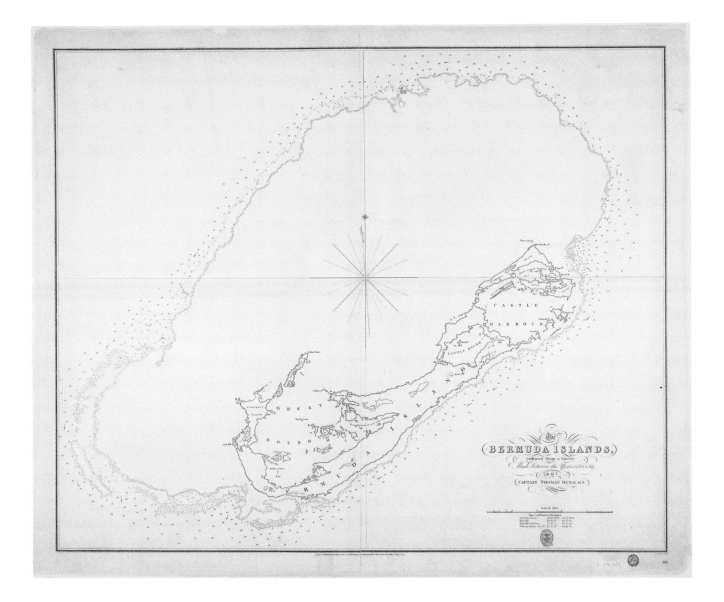

This chart of Bermuda was based on survey work that depended on the expertise and labour of enslaved pilots, including James 'Jemmy' Darrell, Tom Bean and Jacob Pitcairn. At the end of the eighteenth century, Royal Navy officer Thomas Hurd was sent to Bermuda to survey the island, to establish its suitability as a British naval base. Following American independence, the Royal Navy found itself without an operational home in the mid-Atlantic and identifying a suitable site was of huge strategic importance. Bermuda, settled by the English Virginia Company from 1612, had been under Crown jurisdiction since 1684. During the first half of the seventeenth century, enslaved people were taken from the Caribbean to Bermuda to work on tobacco plantations owned by English colonial occupiers. Later that century, the colony pivoted towards maritime commerce, rather than the production of goods, as a source of income and the island's merchant fleet expanded rapidly. Many enslaved Black people began to work at sea and developed sailing and navigational expertise on which the colony's prosperity would depend. By the time Hurd came to survey Bermuda, many of the island's pilots, whose knowledge of local coastal waters enabled vessels to safely enter or leave port, were enslaved people. In this context, Darrell and Pitcairn were able to use the changing military status of the colony – described now as being 'of the utmost importance to protecting Great Britain'[15] – to their advantage. Freed in 1796 because of his work piloting warships, Darrell made use of his considerable reputation to petition successfully for increased pay, as well as increased rights, for free Black people in Bermuda, including the ability to pass on property to descendants.[16]

*The Bermuda Islands reduced from a Survey made between the years 1783 & 97, HM Admiralty, James Darrell, Jacob Pitcairn, Tom Bean, Thomas Hurd, 1827*

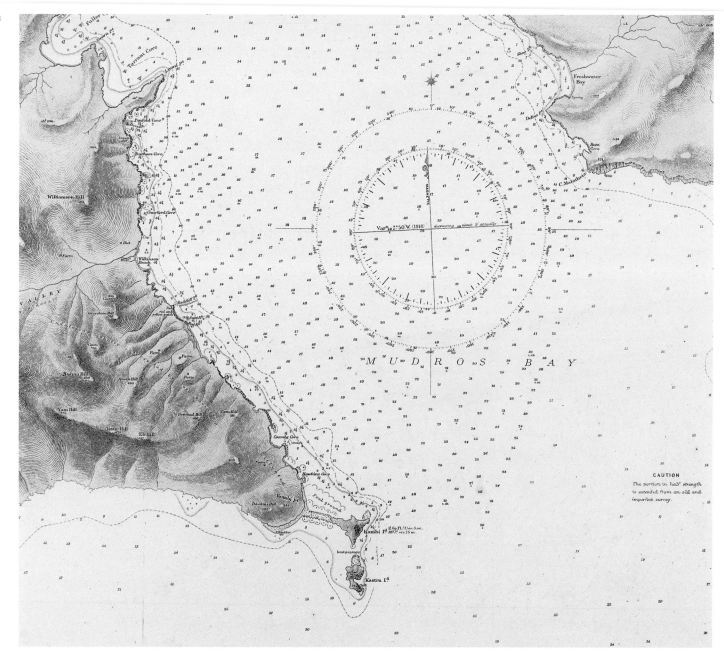

Ensuring that placenames included in surveys are those used locally is an important navigational principle – it is certainly practical to use a single, shared name to refer to a specific geographical point. Historically, however, that principle was one also overlooked in favour of naming places after people – typically officers – on a voyage, or after patrons at home. This was another way in which Indigenous presence was written out of a landscape. Naming could be abused in other ways, too. On an 1890s survey of the Greek island of Lemnos, in the Aegean Sea, Lieutenant Hughes C. Lockyer, who would much rather have been shooting partridges, was ordered instead by his captain, Alvin Corry, to survey the hills around the harbour of Mudros.

Lockyer, not pleased, took the liberty of adding some names as well, inscribing his ire on the coastline, rather than just carving it on his bunk. The hills Yam, Yrroc, Eb and Denmad, spell out, in reverse, the phrase 'May Corry Be Damned'.[17] These names not only made it into the published chart, but were also present in charts and sailing directions published as late as the 1960s.

*Archipelago Lemnos – Port Mudros (Port San Antonio)*, Hydrographic Department, 1918

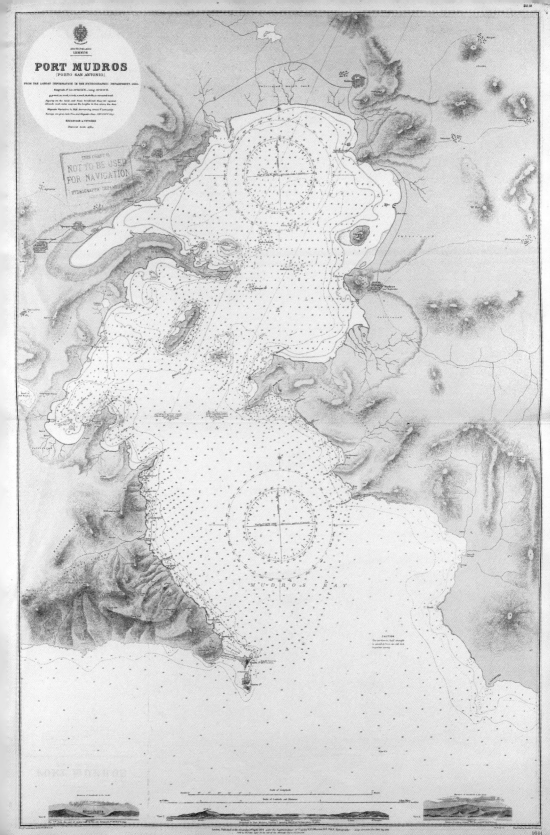

PORT MUDROS
(PORTO SAN ANTONIO)

MUDROS BAY

This chart represents attempts at international institutionalised hydrographic cooperation and the challenges of bringing together and publishing ocean data in the middle of the twentieth century. Part of the third edition of the General Bathymetric Chart of the Ocean (GEBCO), it was published by the International Hydrographic Bureau (IHB), an organisation founded in 1921 to designate international standards for marine charting and to improve the availability of hydrographic information. GEBCO itself had been started in 1903 with the aim of producing a detailed, global picture of the ocean floor. One of the biggest difficulties faced in producing this edition was a vast increase in the amount of available data. In the first half of the twentieth century, the development of echo-sounding technology meant that it was possible to measure ocean depths by sonic pulses bounced off the seabed, rather than by older methods using a weighted line or wire let overboard. Where the first edition of GEBCO used 18,400 soundings, the IHB had to deal with 358,700, far too many to be included on the published sheets. For this sheet, 5,153 (of a total of 37,835 soundings gathered) were included on the published chart.[18] Due to the logistical demands such vast quantities of information placed on the cartographers, as well as the interruption of the Second World War, only 21 out of 24 sheets of the third edition were ever published. The last was issued in 1966, after work on a fourth edition had already begun. But as Julien Thoulet, who developed the GEBCO project, remarked at its outset, 'the Chart is a document which aims continually toward perfection along all its successive publications, but it shall never be finished.'[19]

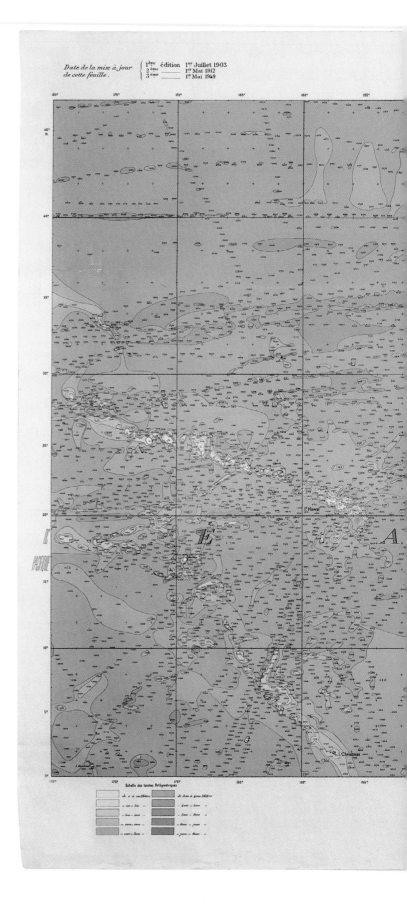

*Carte Générale Bathymétrique des Océans. Sheet A II,* International Hydrographic Bureau, 1949

RTE GÉNÉRALE BATHYMÉTRIQUE DES OCÉANS

LA 2<sup>ème</sup> ÉDITION ONT ÉTÉ PUBLIÉES PAR ORDRE DE S.A.S. LE PRINCE ALBERT I<sup>er</sup> DE MONACO

3<sup>ème</sup> ÉDITION PUBLIÉE PAR LE BUREAU HYDROGRAPHIQUE INTERNATIONAL

Feuille A<sub>II</sub>

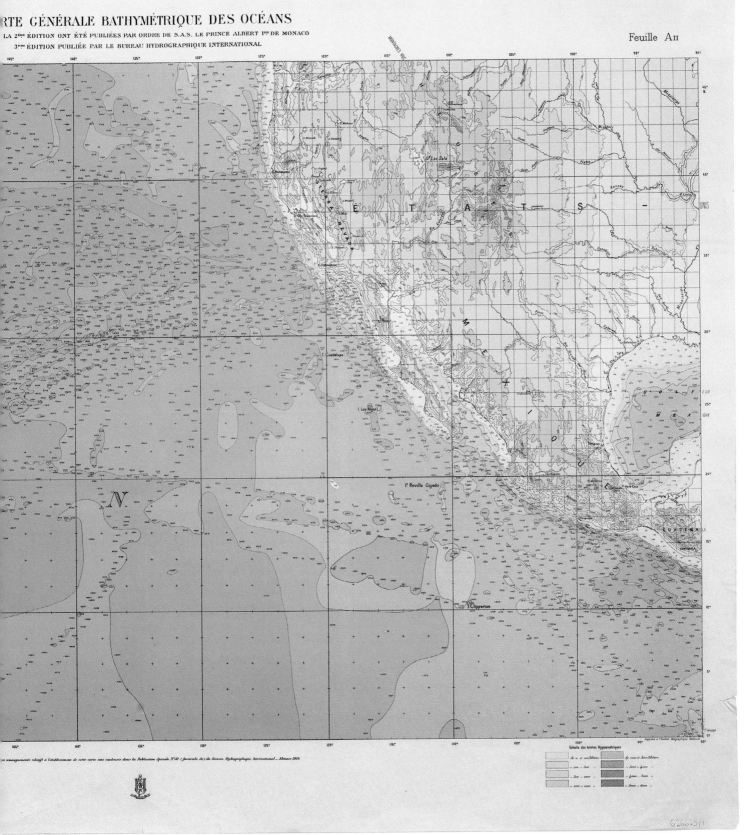

Echelle des teintes Hypsometriques

# I

IS FOR

# INSTRUMENT

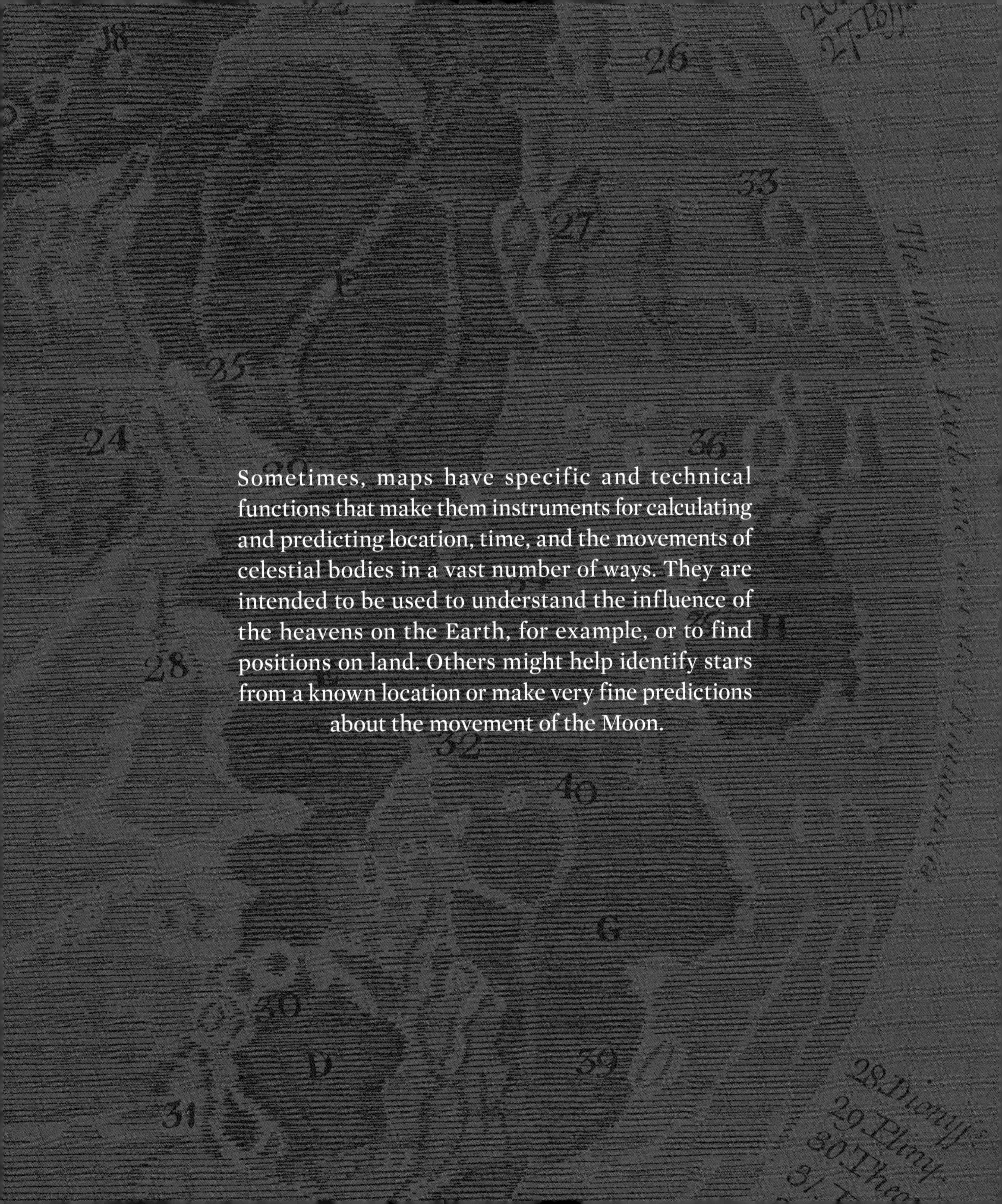

Sometimes, maps have specific and technical functions that make them instruments for calculating and predicting location, time, and the movements of celestial bodies in a vast number of ways. They are intended to be used to understand the influence of the heavens on the Earth, for example, or to find positions on land. Others might help identify stars from a known location or make very fine predictions about the movement of the Moon.

Made in 1537, this celestial globe is the earliest extant to have been printed from copper plates. As well as showing the geography of the stars, celestial globes in early modern Europe were often intended as sophisticated calculating devices. With a graduated horizon ring (horizontal) and meridian ring (vertical), such instruments could be used to predict the time of the rising and setting of stars in different locations and to establish the dates of astrologically significant moments. Contemporary belief in the effects of the movements of the heavens on the Earth meant that astrologers would be consulted about the best times to take significant actions: when to get married, travel or even wage war. The globe was made in Leuven,

in what is now Belgium, by Dutchman Gemma Frisius (1508–55), who tellingly signed himself as '*medicus ac mathematicus*' – doctor and mathematician. The movements of the heavens were also understood to have a direct relation to human health and the stars would be consulted by physicians as they worked out how to treat disease. It has been suggested that it was Frisius's occupation as a doctor which led him to globe-making.[20] It is no surprise, then, that Frisius would emphasise his medical identity on a celestial globe.

Celestial globe, Gemma Frisius, 1537

A MAP OF THE MOON,
with the Names of the dark and light Spots.

In a Lunar Eclipse, the Solar Time when the Shadow touches any of
the Spots, is to be taken as correctly as possible at two distant Places, and the Difference
of those Times is the Difference of Longitude of the two Places. Published by S. Dunn June 14th 1786.

Following the invention of the telescope in the early seventeenth century, astronomers were able to produce detailed maps of the surface of the Moon. Many did so with a particular purpose in mind. Not merely depictions of a world that could never be visited, Moon maps, or selenographies, were instruments which could be used to calculate position on Earth. Specifically, if, during a lunar eclipse, the time at which a particular lunar feature was touched by shadow was recorded from two different locations on Earth, then the difference in longitude (how far east or west a position is) between those

locations could be calculated. This simple map, produced to show the main lunar features, was published in 1786 by Samuel Dunn in his book *The Theory and Practice of Longitude at Sea*. Dunn, based in London, was a teacher of and writer on navigation and in his book he described the different methods of determining longitude. Lunar eclipses were praised as being easy to observe, though Dunn indicated too that because it relied on the occurrence of such an astronomical event, as well as two observers in different places, it was less than practical for navigation.

A simplified celestial globe designed for use in navigation, the Star Finder was invented in the later nineteenth century. It was intended to help identify visible stars while at sea and showed only the positions of the brightest ones. Since the position of particular stars could be used to establish precise location, being able to recognise quickly the stars that were visible at a given time and place was extremely useful. The sphere, set within a box that would protect it from the rolling of the sea, is framed by graduated brass fittings which are used to make calculations. To use the globe, a navigator would adjust the instrument by turning it until their latitude, as marked on the meridian ring (running front to back over the top of the globe and fixed at opposite points), was at the top of the globe. Then they would adjust the instrument for the time, which is inscribed on the celestial equator of the globe itself. The globe would be rotated until the time was underneath the meridian ring. Users would then be able to see which stars were in the sky at their location, based on the stars they could see on the globe. As the pamphlet accompanying the globe explained, 'No one in possession of such a globe could any longer have the excuse that he did not know what star he was taking ... it will hasten the time when the Ship's position will always be taken before dark, and at twilight, by the stars.'

*Star Finder*, Cary & Co., around 1900

By the early twentieth century, data about the Moon's motion was so detailed that understanding apparent irregularities in its orbit became an important area of astronomical research. Scientists came to agree that it was actually the Earth's rotation, not the Moon's orbit, that was inconsistent. Looking to the Moon to understand more about the motion of the Earth, then, astronomers used lunar occultations – that is, the exact moment a star disappears behind the Moon for several minutes – to precisely measure the Moon's position in relation to the background of stars. But trying to predict when these events might occur was no easy task. This object, known as the Occultation Machine, was constructed at the Royal Observatory in Greenwich in 1934 by Leading Joiner A.C.S. Wescott, to predict the timing of lunar occultations as seen from multiple observatories across the world. Using a suspended 12-inch globe, which was turned by a cord, a car headlamp and a lens were used to project both where the Moon's shadow would fall and from where on Earth the light of a particular star could be seen. The point at which the two met represented the predicted time and location of an occultation. Because it relied on observing light on the surface of the globe, the instrument had to be used in as dark a space as possible, which is presumably why the land areas on the globe were painted white in 1950, making the instrument easier to read in low light. The instrument, which could give provisional timings of occultations accurate to one minute, remained in use well into the 1960s.[21]

Occultation Machine,
A.C.S. Wescott, 1934

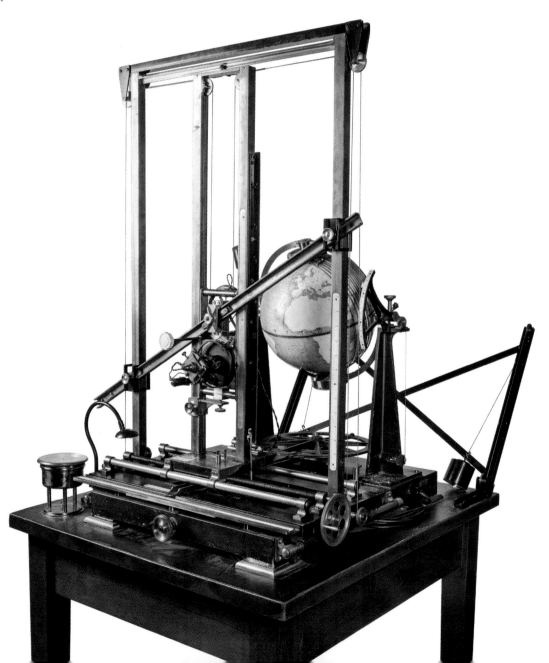

J

IS FOR

JOKE

Cartography is a graphic art and, as such, offers substantial scope for visual play. Showing countries in the form of people (anthropomorphic maps), animals (zoomorphic maps) or using artistic techniques to highlight particular features are all ways in which those involved in making maps experimented with the art form. In doing so, they showed heightened awareness of practices of both drawing and looking. Far from being funny, they often highlighted something deadly serious.

Playing on the metaphors of the 'face' of the Earth and of the human face as a 'map' of character – both current in the sixteenth century – this map, dressed as a court jester, is rather unnerving. It asks the reader to reflect on the foolishness of humanity and the futility of worldly pursuits. The text at the top is the great command of Ancient Greek philosophy: 'Know Thyself'. Elsewhere are suggestions as to what that self-knowledge will reveal. 'The number of fools is infinite', is written at the bottom, while on the cap's ears is the line 'who does not have asses' ears', an allusion to the foolish King Midas of Greek mythology. We are all fools then. But why a map? It fulfils two functions – itself part of the joke. It is a representation of the world. As expressed in a quotation from the Roman philosopher Pliny the Elder, 'this is the world, it is here ... we throw humanity into uproar and launch even civil wars.' For many in sixteenth-century Europe, the world was certainly in uproar, not least because of the violence of the Reformation. On top of that, the map is also a representation of a map and thereby of knowledge as a worldly pursuit. At the top of the jester's rod is the biblical phrase 'Vanity of vanities, all is vanity!'. Mapping itself is mocked, even as it is used.

Fool's head world map,
unknown artist, around 1590

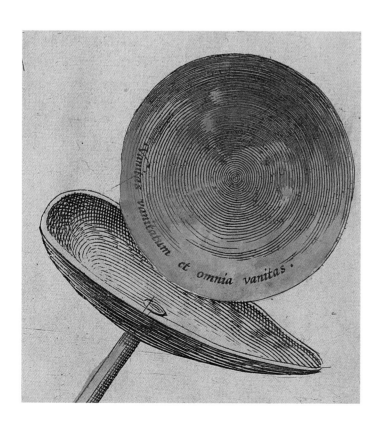

P.305 P

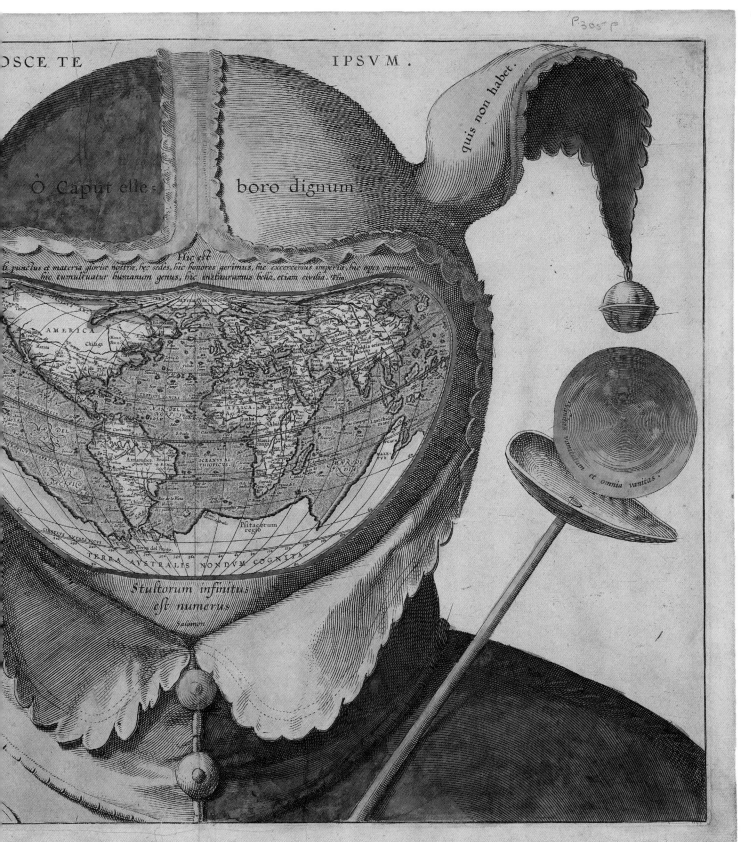

OSCE TE          IPSVM.

quis non habet.

Ô Caput elle= boro dignum

Hic est

...i punctus et materia gloriæ nostræ, hæ sedes, hic honores gerimus, hic excercemus imperia, hic opes...

hic tumultuatur humanum genus, hic instauramus bella, etiam civilia. Tiæ...

Stultorum infinitus

est numerus

Salomon

TERRA AVSTRALIS NONDVM COGNITA

vanitatum et omnia vanitas.

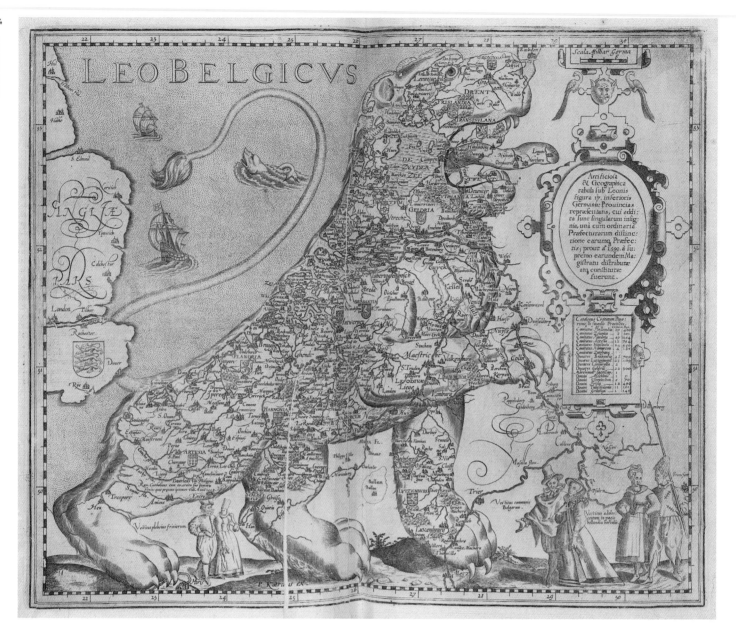

This map brings together the 17 provinces of the Low Countries – today the Netherlands, Belgium and Luxembourg – in the form of a lion. Drawing on the heraldic emblems of these territories, many of which featured lions, and making a point about both unity and might, Leo Belgicus maps emerged during a period of huge upheaval known as the Dutch Revolt or the 80 Years' War (1568–1648). This was a war of self-determination and independence, during which the provinces sought to 'throw off the Spanish yoke', ruled as they were by Philip II of Spain, a member of the House of Habsburg. The seven northern provinces eventually became the Dutch Republic and the ten southern provinces remained under Spanish rule. From 1609, a

truce brought 12 years of peace to the region and the Leo Belgicus remained a symbol of unity and identity. This particular version, first published in 1617, was part of an atlas of the Netherlands, which featured several regional maps, followed by maps of the individual provinces. Produced during a period of rapid economic growth and increasing prosperity in the north, the Leo Belgicus may be visual play, but, as a symbol of regional unity and proud prosperity, it is very serious indeed.

'Leo Belgicus', *La Germanie Inferieure*,
Pieter van den Keere, 1622

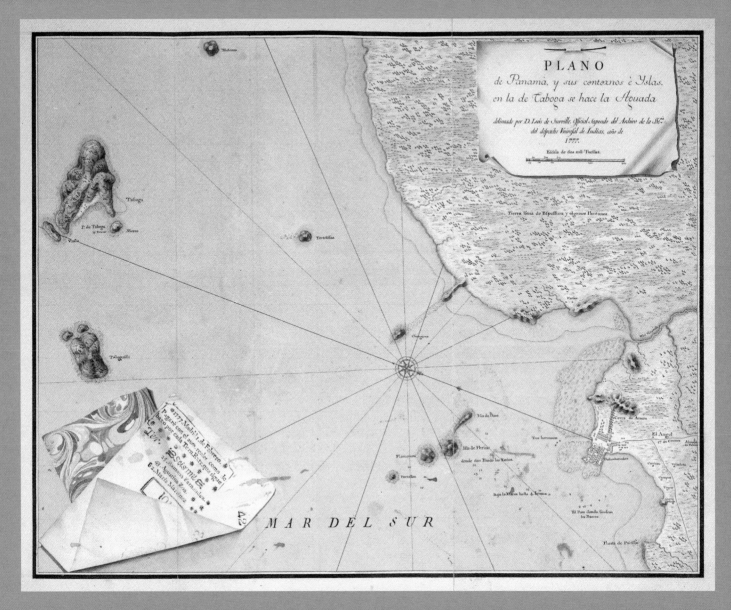

Luis de Surville (active 1770s) went to considerable lengths to demonstrate his graphic skill, producing illusions known as 'trompe l'oeil' (literally 'deceives the eye') to play with the visual and tactile qualities of paper maps. Designed to challenge the eye, trompe l'oeil features on a map highlight the map as a material object, emphasising the active and creative work that goes into making one. Clearly a consummate artist, as Second Officer of the Archive of the Secretariat of State and of the Indies, de Surville was also an archival map-maker. Using the amassed materials of the Spanish colonial archive he produced maps of places he had never visited. Perhaps no wonder, then, that de Surville's visual play was all about paperwork.

The title apparently pinned onto the map points to a standard way of organising documents. The 'casual' placement of scraps of paper in the bottom corner implies a work left for a moment while attending to some other task, with a lottery ticket and piece of marbled paper perhaps absentmindedly removed from a pocket. Such suggestions are, of course, part of the play. The illusion of haphazardness on the map in fact calls attention to the immense skill involved in producing it.

*Plano de Panama, y sus centexnos e Yslas, en la de Tobaga se hace la Aguada*, Luis de Surville, 1777

James Gillray (1756–1815) took anthropomorphic maps to new levels with his depiction of George III as England and Wales. Here, Durham is the king's face and Northumberland his nightcap, Wales is a flapping nightgown and he defecates out of the Solent onto a French invading force. Pontefract, a town famous for making liquorice (used as a laxative), is one of the few English towns marked. Withering, and revelling in a pun, Gillray depicted the French invading force as 'bum boats', that is, small boats which rowed supplies out to ships. Stolidly contemptuous and literally embodying the land of England and Wales, George III is also John Bull, a personification of staunchly anti-Jacobin Englishness. Alongside Gillray's political point-scoring against Napoleonic France is delighted irreverence in drawing attention to the bodily functions of the monarch. He unsubtly questions whether the 'British Declaration', text included in the excrement pouring into the Channel and referring to a vague, non-committal statement made by George III to Parliament in 1793 about the war with France, is, in fact, shit.

*The French Invasion; – or, John Bull bombarding the Bum Boats*, James Gillray, 1851 (first impression 1793)

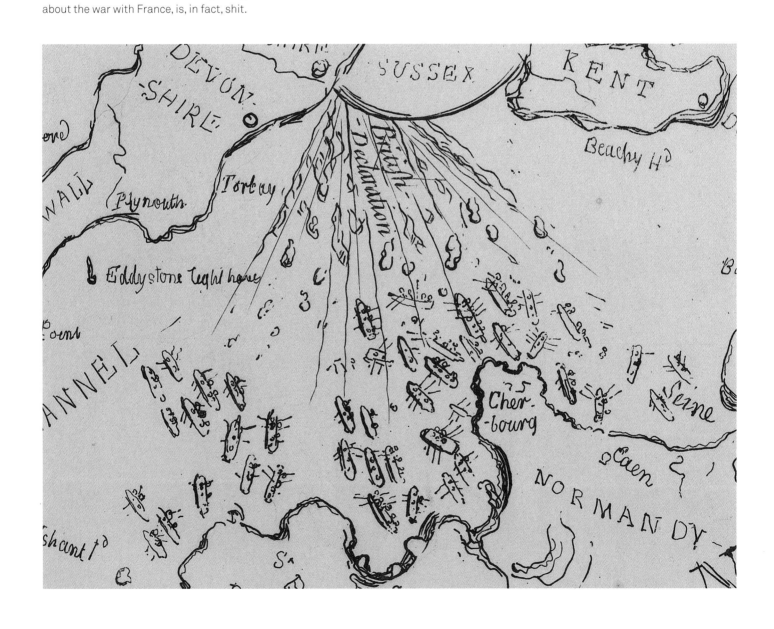

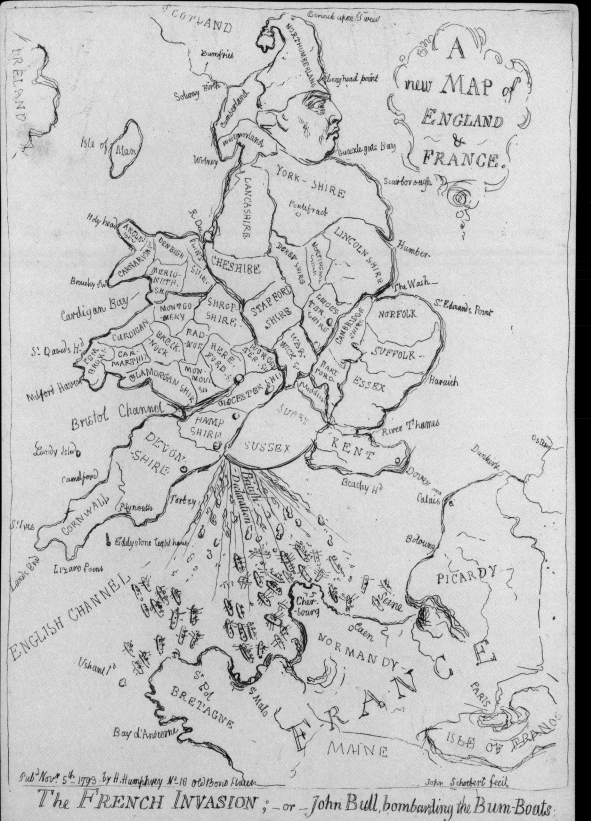

Holborn

Turn-mill Stream

West Smithfield

33

32

Strand

1

2

3

4

3

6

7

8

30

26    29    25

K

IS FOR

KEY

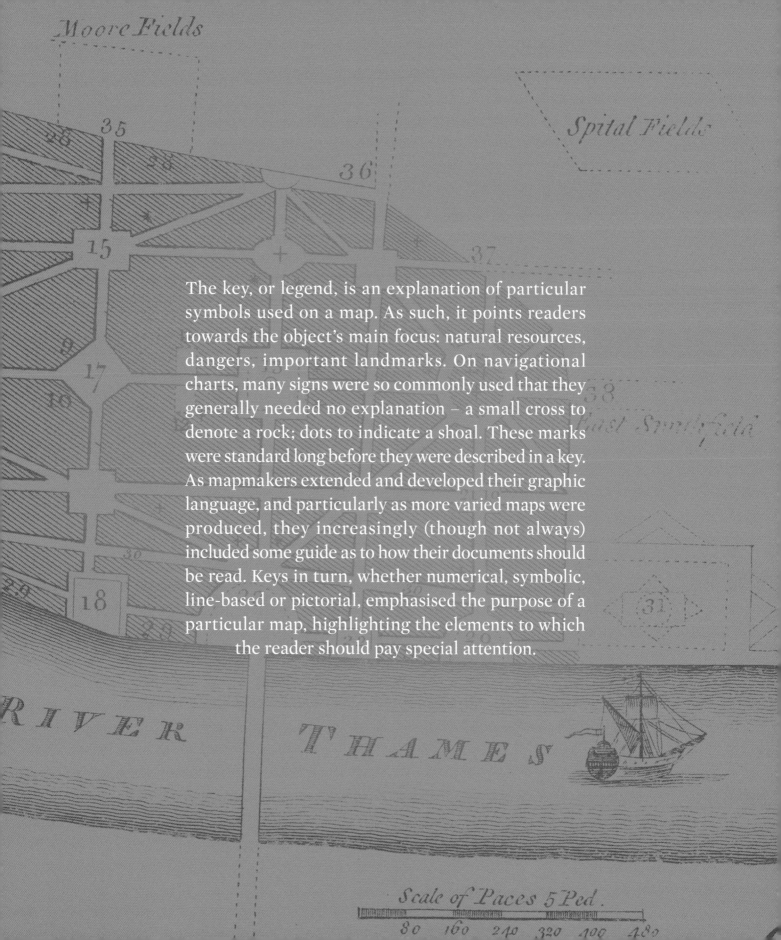

The key, or legend, is an explanation of particular symbols used on a map. As such, it points readers towards the object's main focus: natural resources, dangers, important landmarks. On navigational charts, many signs were so commonly used that they generally needed no explanation – a small cross to denote a rock; dots to indicate a shoal. These marks were standard long before they were described in a key. As mapmakers extended and developed their graphic language, and particularly as more varied maps were produced, they increasingly (though not always) included some guide as to how their documents should be read. Keys in turn, whether numerical, symbolic, line-based or pictorial, emphasised the purpose of a particular map, highlighting the elements to which the reader should pay special attention.

With its gridded streets cut through with a kite shape on one side and an octagon on the other, this new geography of London, proposed in the seventeenth century, drew on contemporary thinking about what the ideal city would look like. In the immediate wake of the Great Fire of London in 1666 that caused huge destruction, King Charles II declared that a 'much more beautiful City' would be constructed from brick and stone and with wider streets to prevent such a calamity from ever happening again. Various proposals were developed, including this one by writer and Fellow of the Royal Society John Evelyn (1620–1706). Neither the planned grid proposed by Evelyn nor a similar scheme suggested by the architect Christopher Wren (1632–1723) were realised and in the eighteenth century the plans became an architectural curiosity, reproduced many times in print. Here, the numbered key enables the reader to make sense of a new and unfamiliar cityscape. Evelyn suggested re-siting parish churches and redrawing parish boundaries, in addition to erecting important civic buildings on the piazzas he arranged at the intersections of different streets. Only the walls and the gates of the city, as well as London Bridge crossing the Thames, remain the same.

*Sir John Evelyn's plan for rebuilding the*
*City of London after the Great Fire in*
*1666, John Evelyn, around 1740*

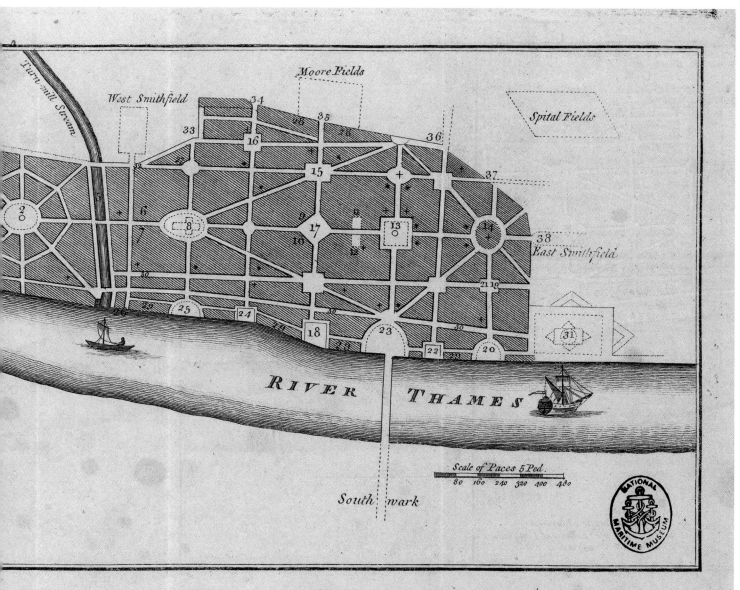

*for Rebuilding the City of London after the Great Fire, in 1666.*

# INDEX CHART

*to the following Survey,*
*of the Coast of*

## KARAMANIA

*and of some other parts of*

## ASIA-MINOR.

by Francis Beaufort F.R.S.
*Captain in the Royal Navy*
1810~1811~1812.

### GENERAL EXPLANATION.

*of the*
Marks *and* Symbols *used in the following* Charts *and* Plans.

* * *

P. *stands for Plan*, V. *for view, &c.*

*The modern names are in Roman characters, and the antient in Italics, in this general Chart*

*Signifies a low Rocky shore*

Rocky Cliffs

*a Sandy beach*

*a Gravel beach*

*a Beach of large loose Stones*

Sand hills, *along the shore*

Loose earthy or marshy shore

*In projecting an extensive Survey, the data will sometimes prove uncertain or suspicious, when any doubt has occurred, the coast is thus expressed in discontinuous lines*

Hills *and* Mountains. *When the acclivity is the same on all sides, the shading is equally dark all round, as in Example 1st But when one part of the shading is narrower and darker, as in Example 2d it is meant to denote that there, the ascent is proportionably steeper. The peak on the summit of Example 3d is on one side a rocky cliff. The figures indicate the perpendicular altitude of the hill, in yards, above the level of the Sea.*

Towns, Villages

Inhabited houses

Deserted buildings, Antient ruins, Theatres &c.

Trees, *Particular clumps only are inserted, in order to avoid confusion, but it is to be understood that the hills and mountains along the coast are generally covered with Firs, and dark evergreen bushes*

One fathom,

Two fathoms,   *These dotted lines show the extent from the shore, or the limits round a shoal, of the several depths of water which are here annexed*

Three fathoms,

Sunken rocks, *with not more than One fathom water on them*

Do   Do   *with not more than Two fathoms water on them*

Do   Do   *with not more than Three fathoms water on them*

*This mark indicates that no bottom was found with the depth of line expressed ⸺ All the soundings are marked in English fathoms. c. denotes a bottom of Clay. cr. Coral rocks. gr. Gravel. gw. Grass and Weeds. m. Mud. r. Rocks. s. Sand. fs, fine sand.*

*An anchor indicates convenient Anchorage for large vessels*

*A grapnel points out the usual anchorage of small coasters, generally near enough to lie with a hawser to the shore*

*An arrow marks the direction of the Current, and the adjacent figure, its rate per hour in miles*

True North

Magnetic North

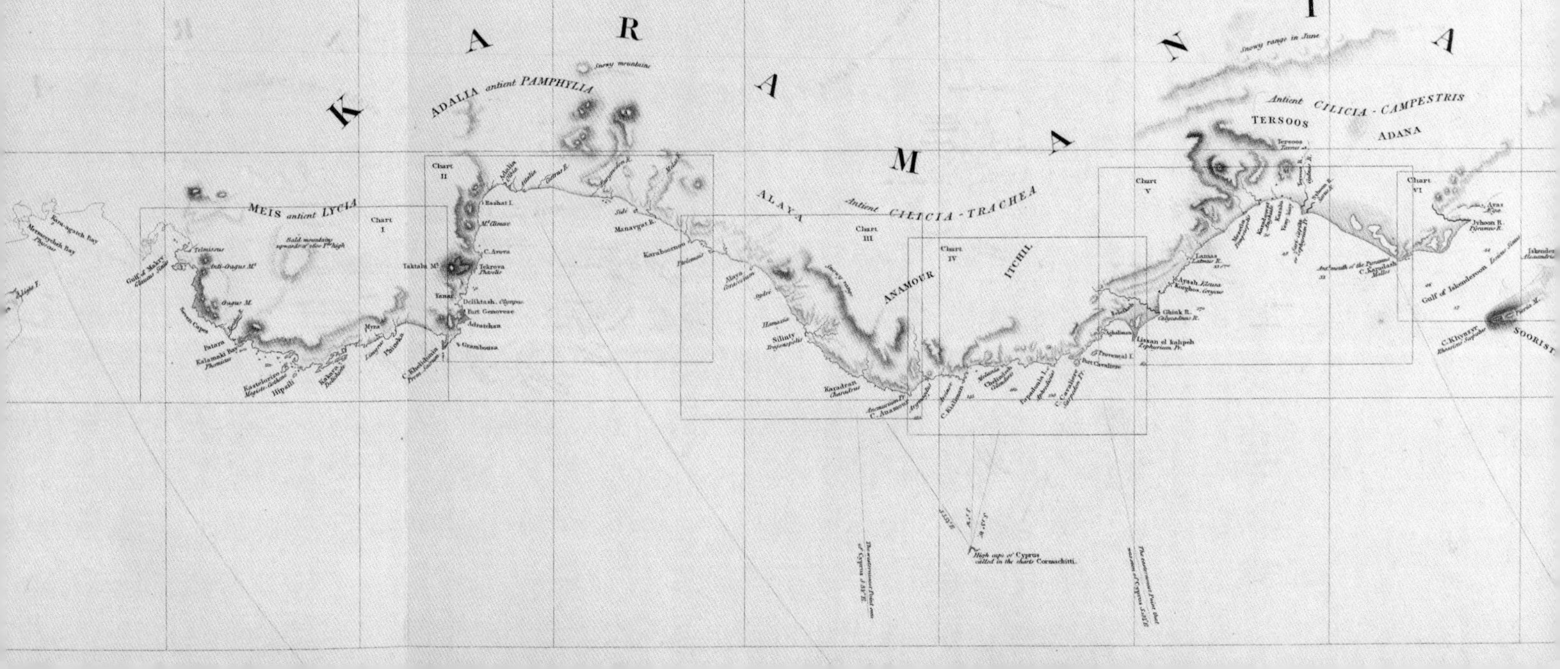

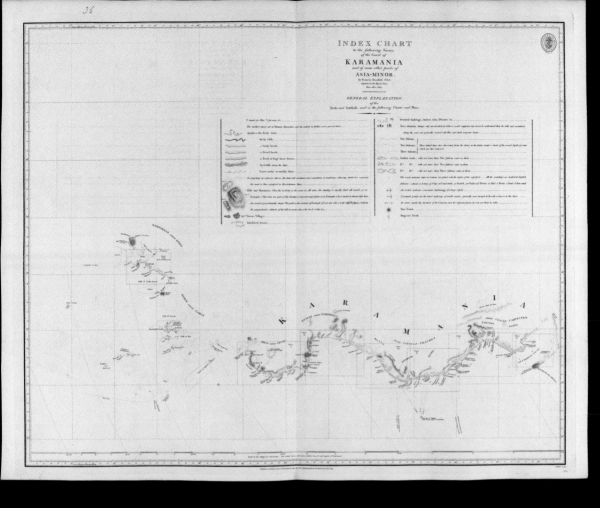

*Index Chart to the Following*
*Survey of the Coast of Karamania*,
Francis Beaufort, 1819

Even before Francis Beaufort (1774–1857) became Hydrographer
to the Navy and did much to standardise the appearance of
Admiralty charts, he thought a lot about symbols, what they
referred to and how to make them understood. His emphasis on
clarity in communication extended beyond his work on charts.
Today, he is perhaps most famous for the wind scale that carries
his name, a system of notation for the strength of wind that was
designed to avoid confusion by giving seafarers shared reference
terms for their description of the weather. Here, on the first sheet
of a set of charts of the Turkish coast, based on a survey carried

out from 1811 to 1812, Beaufort wanted to make absolutely sure
that no one mistook any of the signs he used, no matter how
conventional many of them already were: an anchor indicates good
anchorage; darker shading on a hill references a steeper slope; a
soft line with stipple behind it means a beach is sandy. Musing in
a letter to his father, Daniel Augustus Beaufort, about what to call
such an explanation, Beaufort light-heartedly suggested 'Reading
Made Easy or Hydrographical Language translated for the
Uninitiated or Multum in Parvo, short signs instead of long lines
or Repetition avoided or One Look for All?'[22]

PRINCIPAL ANCHORAGE, WEST COAST, MALDEN ISLAND.

## Sailing Directions for Malden Island.

Ship Masters desirous of making a quick and easy passage from Melbourne to Malden Island, would do well to go through Bank's Straits, and pass to Southward of New Zealand, and not to enter the S.E Trades until the Longitude of the Port is run down, and if not bound to Tahiti, at all times to sight it, leaving Flints Island on the Port side, at this Island and up to the Northward a strong westerly current is felt, and about here the wind draws more from the Northward.

On making Malden Island to the Eastward get 30 Fathoms of Chain ranged before the Windlass, take in and stow all sails, with the exception of Topsails, Jib and Spanker, run along the Coast under these sails, to within a quarter of a mile. The Manager or some other person will come off in a boat and give further Instructions as to anchoring and making fast to the moorings.

Should the Island be sighted from the Westward, never stand to the Northward past the line of the Island; but rather beat up on the South side; the current is very strong to the Westward, to the Northward of the Island.

Latitude 4° 2' South Longitude 154° 50' West.

Reference to View
1  New House and Store
2  Old Quarters
3  New Stone Store
4  Kitchen
5  Kanaker's Quarters
6  Kanaker's Cooking place
7  Condensers and Well
8  Tanks
9  Boats

Scale for Guano Field 4½ inches to the mile
49 Fathoms to the division of the inch

Old Ruins

Guano

Guano Patches

LAGOON

Lagoon

Lagoon

Good Anchorage all along this coast

○ This denotes good Guano
◑ bad having too much lime
◔ good but damp
○ Worked.

Nicholson's Island, nine miles from Station N.E Pt
Anchorage West End Lat 4° 1 S Lon 154° 58' W

## CHART OF
## MALDEN ISLAND.

LAT. 4° 2' SOUTH. LON. 154° 50' W.

Remarkable Trees
in making the land

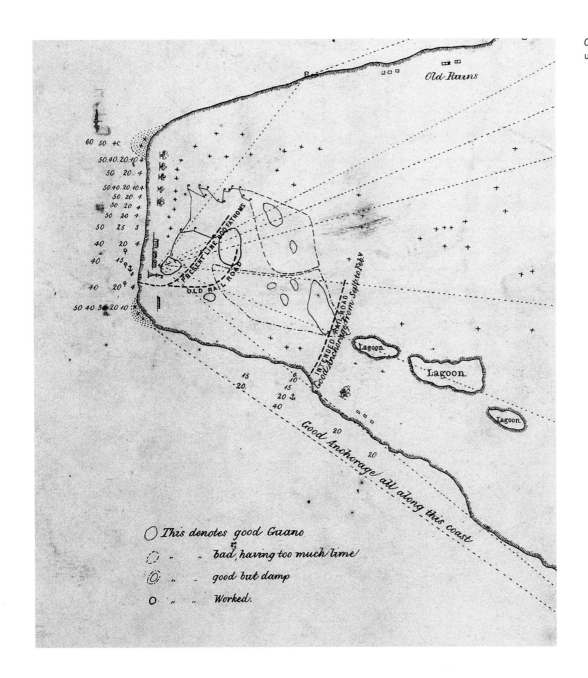

Old Ruins

PRESENT LINE 200 FATHOMS

OLD RAIL ROAD

INTENDED RAIL ROAD

Good Anchorage from Sept to Feb

Lagoon

Lagoon

Lagoon

Good Anchorage all along this coast

◯ This denotes good Guano
◌ " " bad, having too much lime
◎ " " good but damp
○ " " Worked.

In the nineteenth century the Pacific Island of Malden was most famous for one thing: guano. The accumulated excrement of seabirds, guano was extremely valuable as a fertiliser and in the manufacture of gunpowder. It was so prized that in 1856 the US Government passed a law which authorised US citizens to take possession of uninhabited islands rich in the substance, and Peru, Bolivia and Chile fought a war over the location of borders in relation to guano deposits. Little wonder, then, that the key for this map of Malden Island focuses on its quality, using weighted lines to differentiate between deposits. A Melbourne-based company spent the equivalent of around £1 million building infrastructure, including a railway, to support the

extraction of guano. People from the islands of Niue and Aitutaki, almost 1,500 miles to the southwest and about 1,000 miles to the south of Malden respectively, were employed there as labourers on one- or two-year contracts, doing work that often significantly damaged their health. In 1870, a travel writer who visited Malden reported that those digging guano were paid £2 (£195 today) a month, half of which they were required to spend in the company store on the island.[23] In 1907, it was estimated that guano to the value of £500,000 (£54 million today) had come from Malden to be sold around the world.[24] While the key for this map is small and simple, the commodity it described, and the labour that extracted it, was anything but.

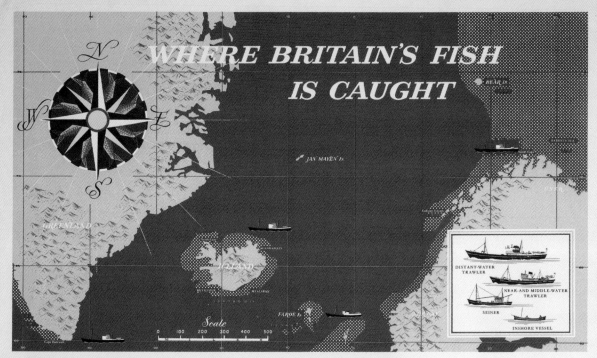

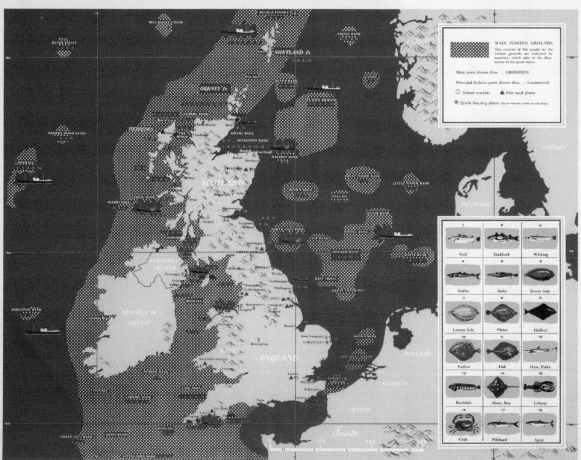

In the decades after the Second World War, there were significant worries about the future of fishing in Britain. The White Fish Authority (WFA) was formed in 1951 to regulate, reorganise and develop the white-fish industry. One part of its work was to promote the benefits of eating fish to the British public, whose consumption of it was falling. The WFA used various measures to do this. Names of fish were changed to make them more appealing: coal fish, for example, became coley. Newspapers carried recipes introduced with terrible puns. A leaflet, 'Fish: From the Sea to the Table', set out to persuade readers that 'the story of fish is an exciting one' and pointed those who wanted to learn more to wall maps like this. Such maps were intended for use by teachers and showed the geography of the fishing industry in Britain, providing information about where different fish came from, what they looked like and where they were processed and sold.[25] Increased familiarity, it was hoped, would make fish more appealing to shoppers. In the 1960s, the WFA specifically aimed their material at women and girls, typically responsible for household purchases. The campaign, though, was not a success and those who worked in the industry viewed it with disdain.[26] Nonetheless, this map, with its pictorial key, was part of the attempt to 'get the people of this country fish-minded enough to eat good quality fish'.[27]

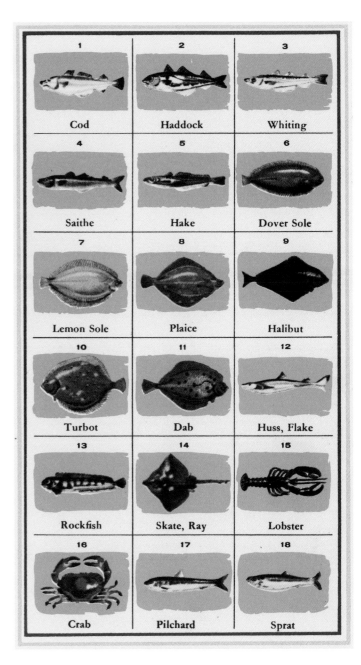

*Where Britain's Fish is Caught*,
White Fish Authority, 1968

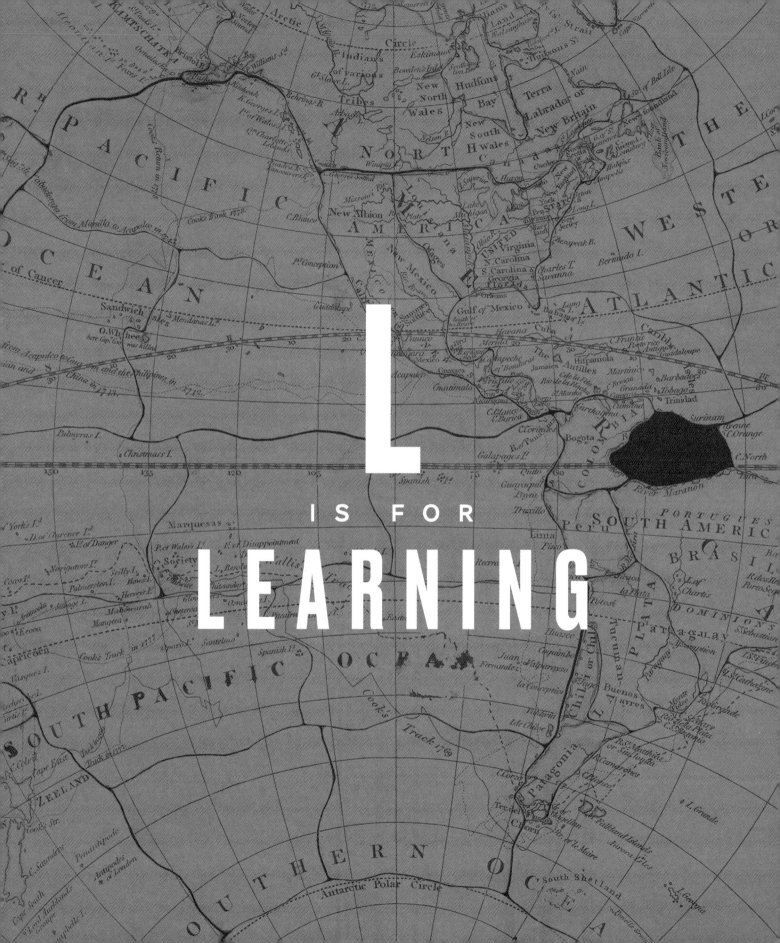

L

IS FOR

LEARNING

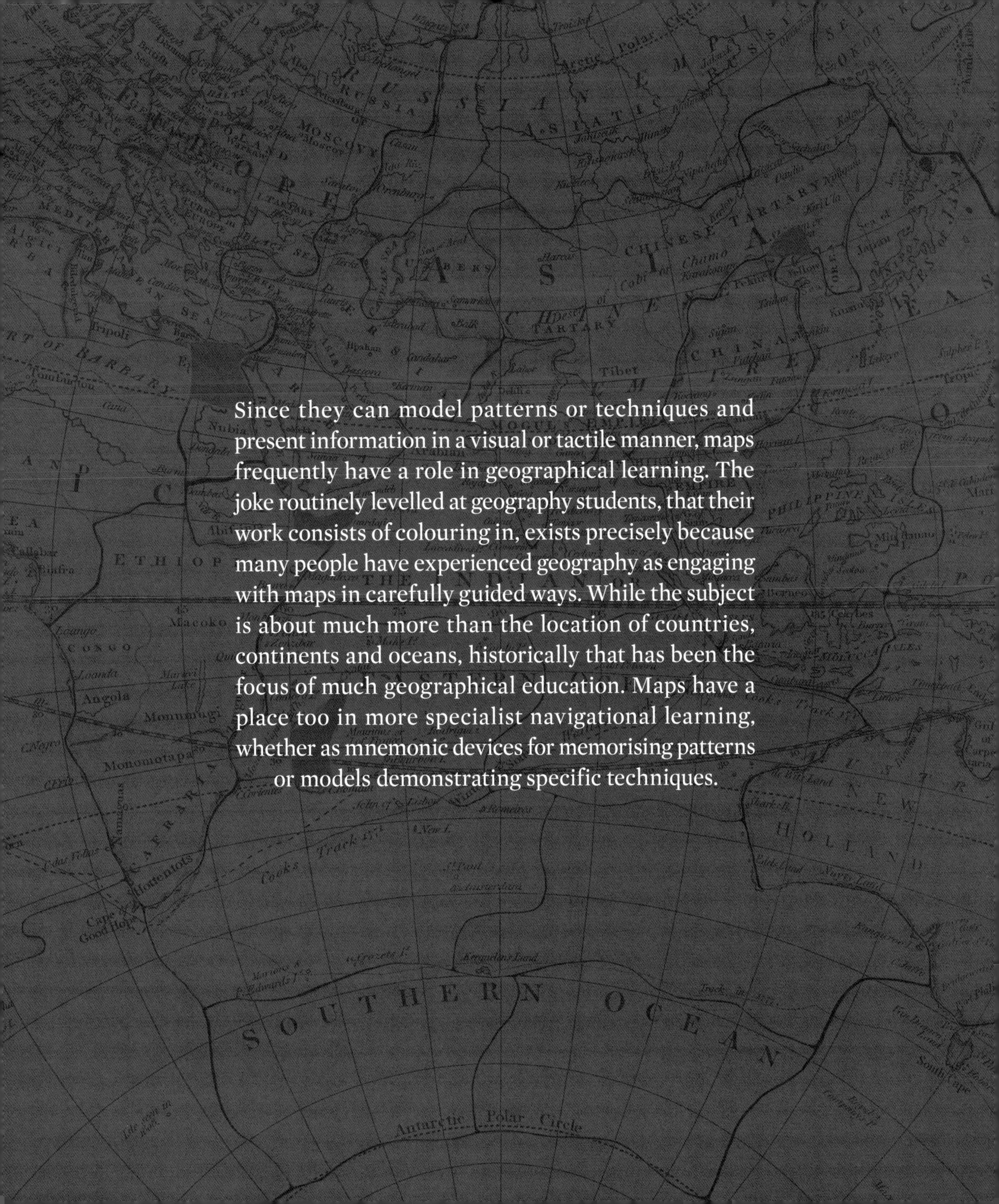

Since they can model patterns or techniques and present information in a visual or tactile manner, maps frequently have a role in geographical learning. The joke routinely levelled at geography students, that their work consists of colouring in, exists precisely because many people have experienced geography as engaging with maps in carefully guided ways. While the subject is about much more than the location of countries, continents and oceans, historically that has been the focus of much geographical education. Maps have a place too in more specialist navigational learning, whether as mnemonic devices for memorising patterns or models demonstrating specific techniques.

In Britain and North America in the decades around 1800, embroidering maps was understood as an important part of a young lady's geographical education. At a time when copying maps was described as 'the ONLY means by which an acquaintance with geography can be perfected',[28] making such maps combined and developed two sorts of genteel social accomplishment: needleworking skill and geographical learning. Map embroidery quickly became popular enough that publishers began to print special templates on paper and directly on satin. We do not know whether this example was made using an embroidery template or traced from a standard map. The geographical knowledge gained was the arrangement of countries, continents, oceans and seas, as well as what was called 'mathematical geography'. This was understood as the framework of lines used to describe the globe: latitude and longitude; the Equator; the Tropics of Cancer and Capricorn; the Arctic and Antarctic Circles. This map is also surrounded by flower sprays, a decorative detail that allowed makers to develop and demonstrate individual flair. In various places the silk has split as the fibres have become brittle, typically the result of long exposure to light. The damage is perhaps indicative of long display in its domestic context, an indication of pride in the painstakingly made map.

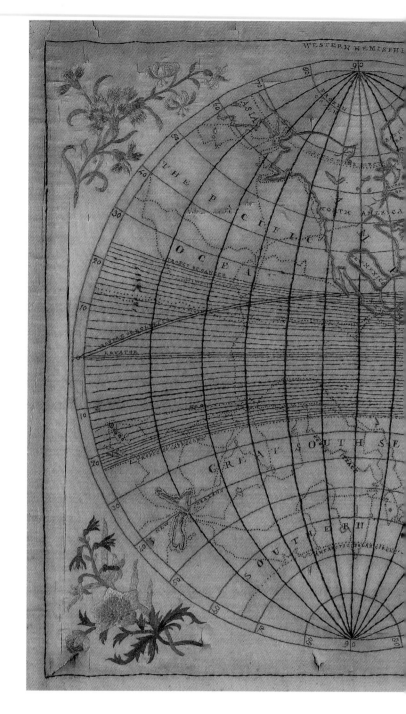

*A Map of the World*, unknown maker, around 1805

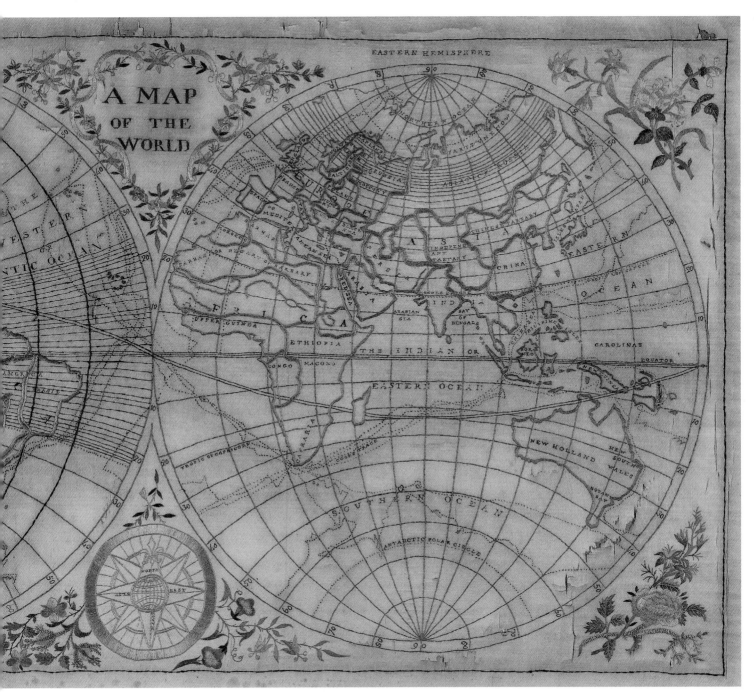

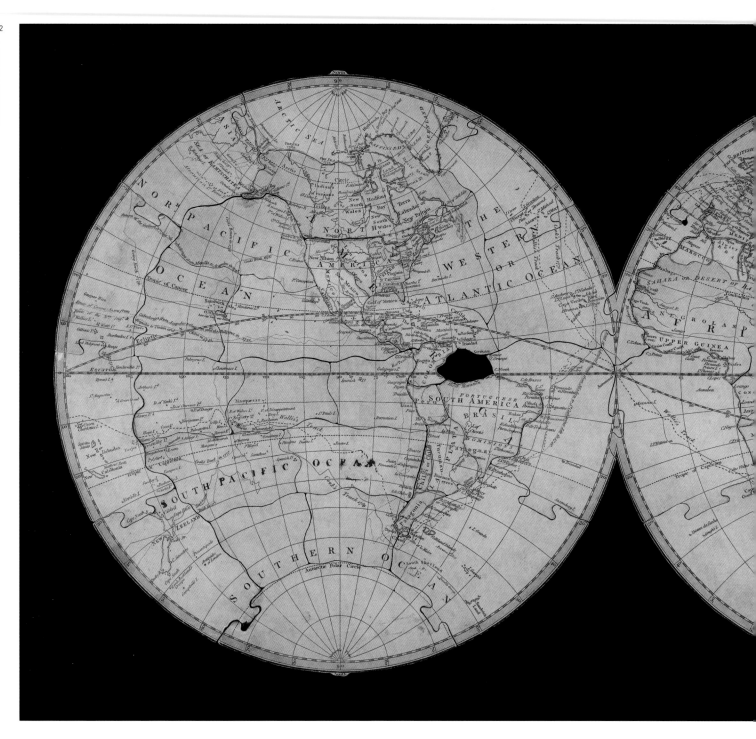

It was in the eighteenth century that people in Europe began to take seriously the idea that play was a good way for young people to learn. As a result, new educational toys and games were developed, including what were known as 'dissected maps', first produced in the 1760s. The first form of jigsaw puzzle, dissected maps became very popular geographical learning aids in well-off families. Indeed, in Jane Austen's 1814 novel *Mansfield Park*, the poor cousin of a wealthy family is mocked, because 'she cannot put the map of Europe together', her relatives mistaking lack of opportunity for ignorance. These toys did not come cheap. Until the 1830s, hand-coloured prints like this one tended to be pasted onto mahogany, by then a valuable hardwood, before being cut into jigsaw shapes by hand. This points not only to the exclusive luxury of geographical toys in the earlier nineteenth century, but also to global systems

*A New Map of the World according
to the Latest Discoveries,* unknown
maker, after 1821

of colonialism: mahogany was harvested by enslaved people
in appalling conditions in the Caribbean and Central America.
The world map therefore encourages an understanding of world
geography that emphasises imperial identity by depicting British
'discoveries' and voyages, including those of James Cook and George
Anson to the Pacific in the eighteenth century. At the same time, its
material existence relied on the very systems it describes.

This geographical skeleton reduces the surface of the Earth to a mathematical framework. Lines of longitude are made from curved brass tubes, latitude from wires and the Equator and the Greenwich Meridian from strips of ivory. A cord at the South Pole is connected to the wooden base, from where the globe can be turned by means of a concealed handle. Across the globe's surface are a variety of triangles for calculation and tracks adorned with tiny ivory ship models. The globe was designed to teach plane sailing, a navigational technique that uses the assumption that the world is flat, with lines of longitude parallel to each other, and thereby avoids the more complicated mathematical work of spherical trigonometry. As a way of determining short distances, plane sailing – with its more straightforward calculations – had much to recommend it and learning the method was an important part of navigational education in the nineteenth century. Charles Hatch, who designed the globe, ran a 'Classical, Commercial and Naval Academy' in Teignmouth in Devon, the sort of private educational establishment that taught navigation at a time when so many livelihoods were connected to the sea.

Nautical globe, Charles Hatch, 1854

Stick chart, unknown
maker, 1970

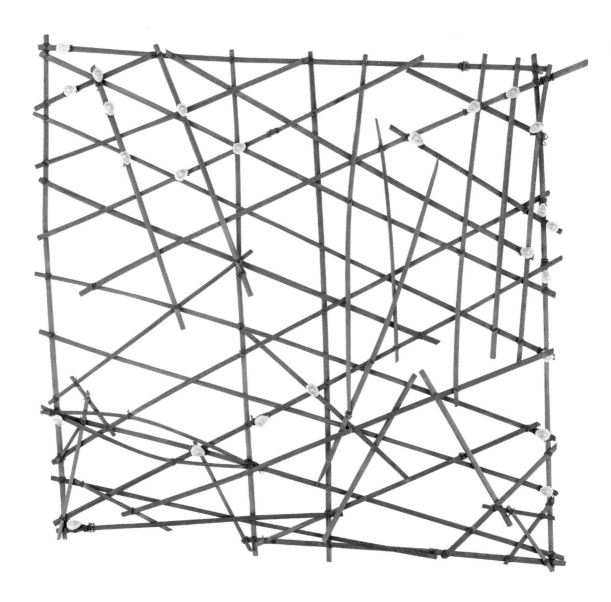

This stick chart, made in the Marshall Islands in the western Pacific Ocean, depicts ocean swells and islands using wooden strips and cowrie shells. The Marshall Islands are a group of 29 coral atolls and 5 coral islands just north of the Equator. In two main chains, they extend over 575 miles. For thousands of years Marshallese seafarers travelled between them and across vast stretches of water, using a technique known as wave piloting, which involves judging location and direction by the feel of the ocean. Young people learning to navigate would spend hours at sea blindfolded, becoming familiar with the minute shifts in the feel of swells in different regions. Alongside such exercises went shore-based learning, where, from the nineteenth century at least, stick charts like this one would help navigators learn and memorise the patterns of the ocean.[29] In the nineteenth and twentieth centuries, German and Japanese colonial prevention of inter-island canoe travel, a systematic devaluing of traditional knowledge and later US nuclear testing and subsequent forced relocation meant that knowledge of these sophisticated navigational techniques fell into decline. In the second half of the twentieth century, Islanders have revitalised and developed traditional techniques as part of a cultural revival that is happening across the Pacific. This particular chart was made for the tourist trade in the 1970s and speaks to the development of tourist economies in the Pacific, as well as the resurgence of sophisticated navigational practices and thus to Marshallese resilience.

M
IS FOR
MANUSCRIPT

Maps, charts and globes drawn by hand are often peculiarly compelling. Medieval works sit within a purely manuscript tradition in which documents would be copied and recopied by hand, sometimes over hundreds of years. When, particularly from the sixteenth century onwards, printed maps began to circulate more widely, far from replacing manuscript traditions, they existed alongside them. Prints were worked up from manuscripts; manuscripts were copied from prints. Sometimes artists attempted to make one form look like the other; sometimes maps contain both printed and manuscript elements. And whether the result of pen or pencil marks of a draft in construction, or the fine illuminated work of a skilled miniaturist, each one is unique.

This map, and the volume to which it belongs, is part of a textual tradition of geographical manuscripts that lasted for almost a millennium. The work was authored in the rich literary atmosphere of the medieval Islamic world by tenth-century scholar Abu Ishaq Ibrahim ibn Muhammad al-Farisi al-Istakhri. Al-Istakhri described the different regions ruled over by the Abbasid Caliphate, including details of major cities, regional customs and produce, and tax revenue. He detailed 'individual pictures [that] show the outlines of each region in the Islamic world, the cities it encompasses and everything that needs to be known about it'.[30] Copies of the original

were made over many centuries and distinctive versions have emerged. Different scribes introduced different artistic flourishes and elements to personalise their work. In this map of Mesopotamia, a historic region that encompassed much of modern-day Iraq, Syria, Kuwait, Iran and Turkey, the mountain ranges – in red – are decorated with floral or geometric patterns. Two rivers are shown in blue: the curved line to the right represents the Tigris; the straight line to the left the Euphrates. This particular volume is one of the oldest of 59 known copies in the world today, with a handwritten statement of ownership inside which dates it to before 1282.

Mesopotamia,
Abu Ishaq Ibrahim ibn
Muhammad al-Farisi
al-Istakhri, before
1282

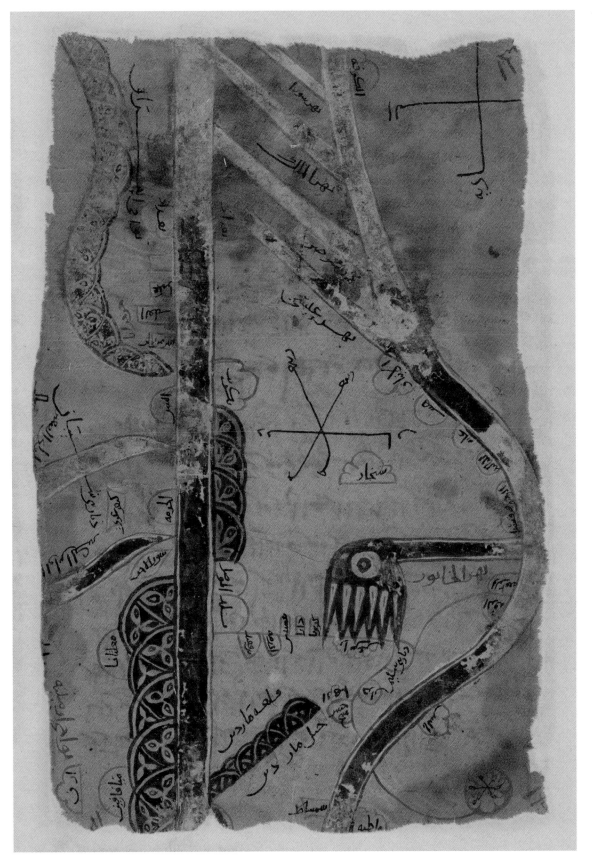

MAISTRO·NORWEST.

TROPICVS·CANCRI.

PONENTE·WEST.

ASIA

mare maius

candia

AEGYPTVS

A·ETHIOP

TA·SVB·AEGY

P·TO.

babil

damasc

ARABIA
mecha

alexandria

sanctaelona

insula s.laurentij

CAPO·DE·BONA·SPERANZ

60
50
40
30
20
10
10
20
30
40
50

GREGO·NORDOST·

china·ciuitaf·magna

SOPHI·

R·DECABA
IA·
cabana

P·DENARS
INGA·
bengala
pegu

SINVSMAGNVS

loflatones

ilabana

calec
ut

bifnagua·done
fcat·u·adiamāti

gilolo

cattigara·finarū

ceilan

LEVANTE·OST·

100    110         120          130         190        160       170        180

AEQVINOCTIALIS·  tap:obama
inſula

iaua

iaua

el·caftullo·de·timo:
de·portogall

banco·de·patron

nazare
piloto
ma

TROPICVS·CAPRICORNI·

INDICVM·MARE·

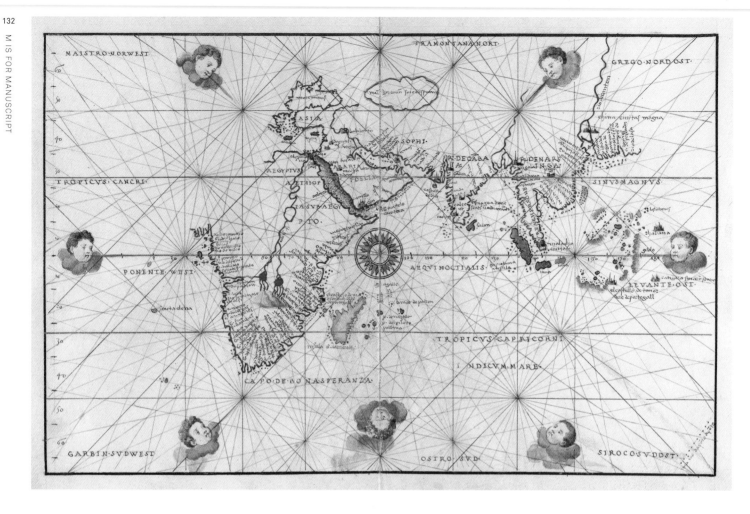

Made in Venice in 1555 by prolific manuscript cartographer Battista Agnese (c.1500–64), this ornate map of the Indian Ocean turns once practical features of early navigational charts into sumptuous decoration. The criss-crossing network of lines, known as rhumb lines, are useful for establishing compass bearings and coloured for easy interpretation at sea. Here, they are an ornamental nod to the navigational practices that had brought such wealth to the Venetian Republic. There are 95 known copies of Agnese manuscript volumes in the world today; such an abundance of examples has allowed historians to understand how his work changed over time, incorporating newly published information and the latest geographical knowledge. Different handwriting and finishes on the various manuscripts, as well as the sheer number of volumes produced, suggest that they were made in a workshop. Not much more is known, as the volumes that bear Agnese's name or are made in this distinctive style are the only evidence we have of him or his workshop. Clearly made for a wealthy client, the volume's original binding also features a practical tool turned into luxury item: into the backboard is set a small, functional compass – a finely crafted, playful curiosity augmenting the work.

Map of the Indian Ocean, Battista Agnese, 1555

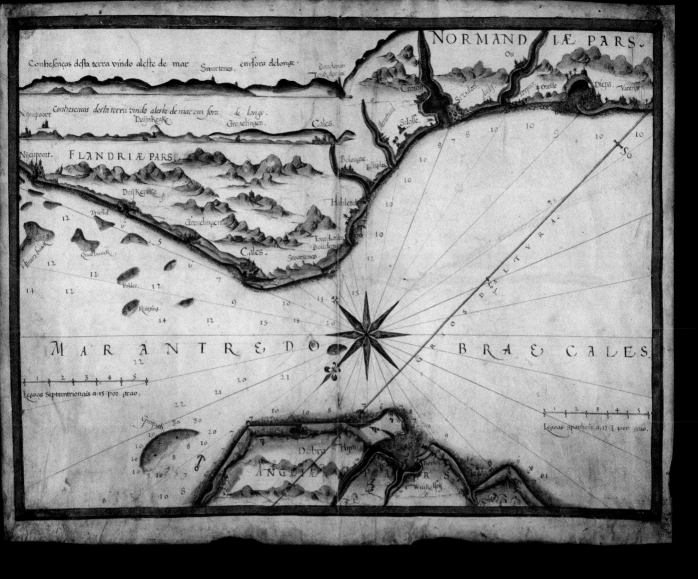

This chart, which shows the Channel between Dover and Calais, was copied from a printed atlas at the end of the sixteenth century. A Portuguese translation of the fourth sheet of Dutch cartographer and navigator Lucas Janszoon Waghenaer's *Spieghel der Zeevaerdt* (1584), it prompts us to think about the complex relationship between manuscript and print. Waghenaer's work, now famous as the first printed sea atlas, drew on his experiences as a mariner and his exposure to Portuguese, Spanish and Italian manuscript navigational aids, including charts. He was also influenced by the tools and techniques of northern European seafarers, who tended to rely on textual descriptions of maritime routes. Describing – in charts, text and coastal views – the coasts along the route from the North Sea to the Mediterranean, the atlas was a great success and was reprinted and translated into several languages. Publishers also came to use it as a model for later sea atlases. Its focus was explicitly navigational, as demonstrated by

the fact that the details of harbours are enlarged, while coastlines where vessels would not seek to anchor have been compressed. This particular chart, carefully copied, translates the better known places into Portuguese – 'Doueren' (Dover) becomes 'Dobra'; 'Diepe' (Dieppe), 'Diepa'. Other smaller or less frequented places retain their Dutch names. Winchelsea remains 'Winkelseij' and Dunkirk is 'Duijnkerke'. At the same time, the map-maker adjusted their version of the chart by adding a latitude scale, the thick red diagonal to the right-hand side. This had been a feature of Portuguese charts since the early sixteenth century, relevant for astronomical methods of navigation less commonly used by northern mariners. For Portuguese users, the printed chart was brought up to date by its reproduction in manuscript.

*Mar antre Dobra e Cales*,
**attributed to Luis Teixera, 1587**

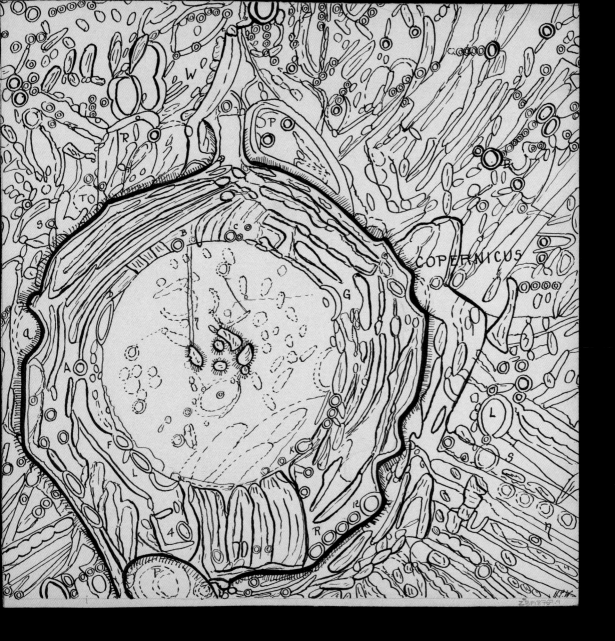

In 1946, amateur astronomer Hugh Percy Wilkins (1896–1960) produced the most detailed map of the Moon published up until that point. Drawn at 300 inches diameter, it was printed in reduced form at 100 inches (2.54 m). This manuscript of the crater Copernicus was just one part of the work of drawing and compiling that went into making Wilkins' map of the entire visible lunar surface. As Director of the British Astronomical Association's Lunar Section, Wilkins developed an international network of correspondents, visited major observatories and worked with his own telescope from his garden in Bexleyheath, south London. He made detailed drawings of light and shadow as they changed on the Moon's surface in order, eventually, to determine the shape of a particular feature and represent it in

outline as a crater, peak, or valley. As Wilkins described, 'much of the finest detail, the most difficult to depict, but by far the most fascinating, vanishes before one's eyes, dying, so to speak, even while it is born.'[31] Compiling his own and others' observations at home by hand, Wilkins then checked it all with further time at his telescope. His *Map of the Moon* was reproduced photolithographically for publication, a process that ensured the distinctive style of his drawing was retained in the finished map.

*Copernicus*, Hugh Percy Wilkins,
about 1945

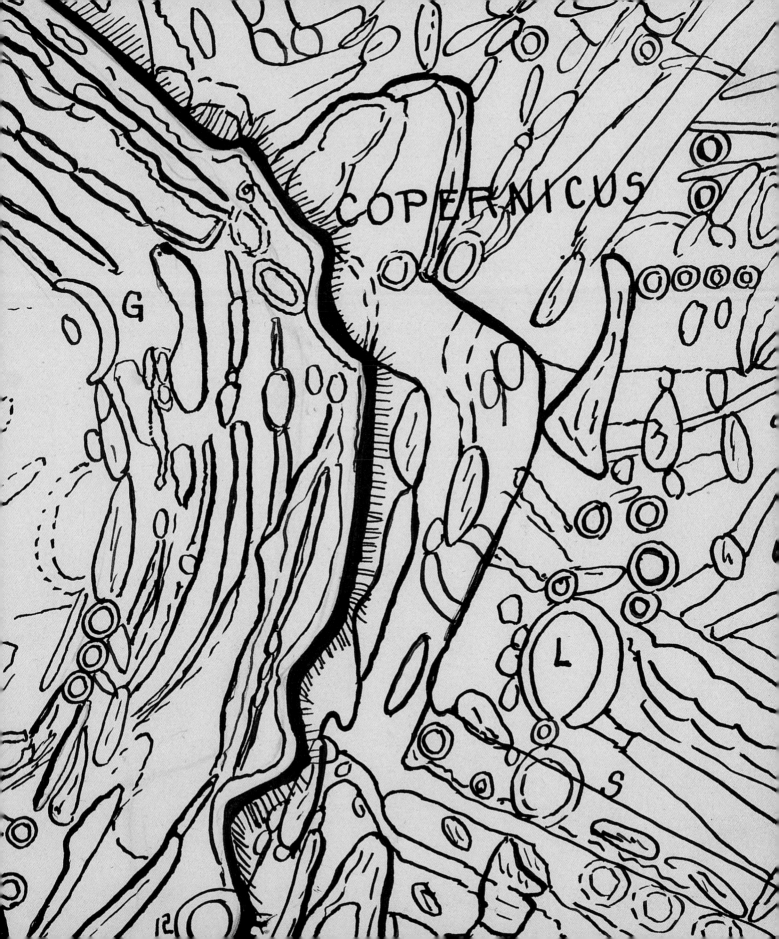

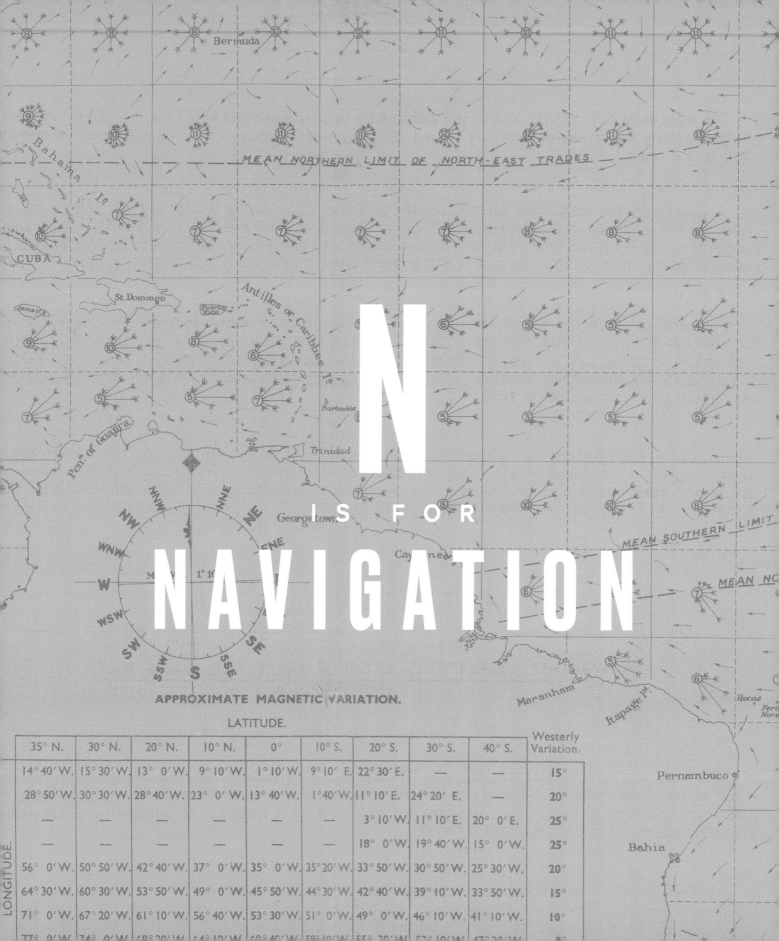

# N

IS FOR

# NAVIGATION

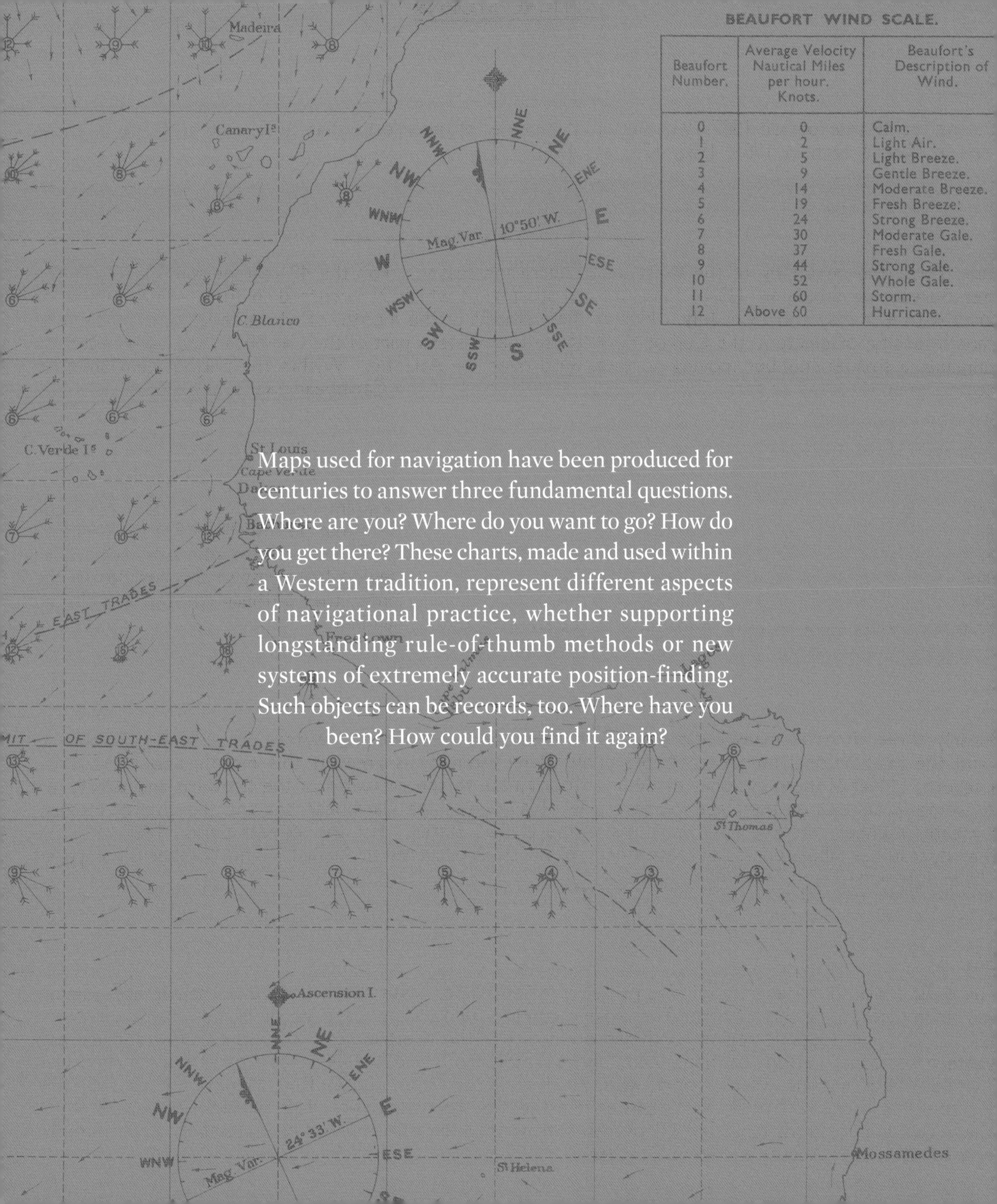

BEAUFORT WIND SCALE.
Beaufort Number.
Average Velocity Nautical Miles per hour. Knots.
Beaufort's Description of Wind.
0 0 Calm.
1 2 Light Air.
2 5 Light Breeze.
3 9 Gentle Breeze.
4 14 Moderate Breeze.
5 19 Fresh Breeze.
6 24 Strong Breeze.
7 30 Moderate Gale.
8 37 Fresh Gale.
9 44 Strong Gale.
10 52 Whole Gale.
11 60 Storm.
12 Above 60 Hurricane.
Madeira
Canary Is
C. Blanco
C. Verde Is
St Louis
Mag. Var. 10°50' W.
EAST TRADES
OF SOUTH-EAST TRADES
St Thomas
Ascension I.
Mag. Var. 24°33' W.
St Helena
Mossamedes

Maps used for navigation have been produced for centuries to answer three fundamental questions. Where are you? Where do you want to go? How do you get there? These charts, made and used within a Western tradition, represent different aspects of navigational practice, whether supporting longstanding rule-of-thumb methods or new systems of extremely accurate position-finding. Such objects can be records, too. Where have you been? How could you find it again?

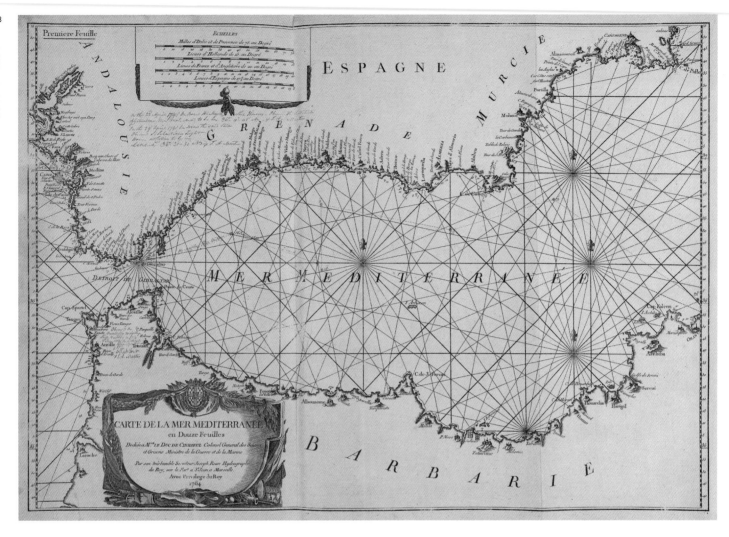

In May 1803, England declared war on France, and Royal Navy vessels proceeded to blockade French harbours in the Mediterranean. One of those vessels was Vice-Admiral Horatio Nelson's flagship, HMS *Victory*. The annotations on this chart, one sheet of Joseph Roux's 1764 *Carte de la mer mediterranée*, show that it was used on board. The chart has the *Victory*'s track eastwards from Gibraltar into the Mediterranean, which was drawn by Thomas Atkinson, who, as the ship's Master, was responsible for navigation. Atkinson was a highly skilled navigator, whom Nelson had expressly requested, and described as 'one of the best Masters I have seen in the Royal Navy'.[32] The points on the track, labelled with dates (16, 17 and 18 June) are the positions from noon observations. This is when latitude is typically calculated, as the Sun is at its highest and appears to 'hang' in the air. The course of the ship is shown as a straight line between these locations. As well as navigating

vessels, Masters were also expected to make observations, correct positional data and carry out surveys, which would be submitted to the Admiralty. Atkinson did this diligently: on this sheet he has noted the results of an observation of the latitude of Cape Spartel in North Africa, significant because of its location at the Atlantic entrance to the Strait of Gibraltar. We do not know exactly why Atkinson used a volume of French charts published in 1764, almost 40 years earlier, to plot *Victory*'s track and record his observations. We do, however, know that in Atkinson's hands, the volume – annotated and supplemented – was a navigational work in progress.

*Carte de la mer mediterranée: première feuille,*
Joseph Roux, 1760, with annotations by
Thomas Atkinson, 1803

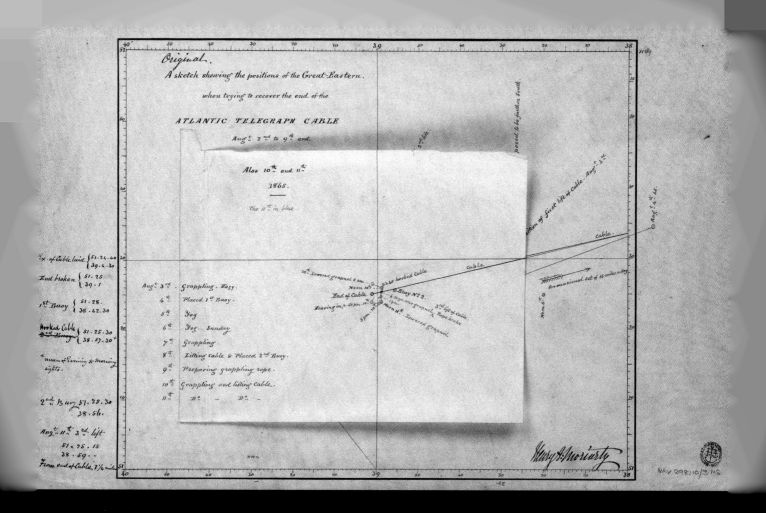

*A sketch shewing the position of the 'Great Eastern' when trying to recover the end of the Atlantic telegraph cable*, Henry Moriarty, 1865

Trying to lift a broken cable from around 2 miles below the surface of the sea in the middle of the Atlantic Ocean seems like a near-impossible task. But, in August 1865, it was precisely this that the crew of the SS *Great Eastern* attempted. The ship – the biggest of its day at 211 metres long (the length of 21 double decker buses) – was laying a submarine cable between Valentia in Ireland and Heart's Content, Newfoundland, Canada. Connecting Britain and North America by electric signal was a heady prospect, especially for those involved in imperial and financial communications. A cable laid in 1858 had lasted only 27 days and there was high excitement about the 1865 expedition. Newspaper coverage on both sides of the Atlantic was extensive and those on board departed well-prepared to make history, intent on recording events and making diaries of what they hoped would be a momentous voyage. The cable, however, broke

This chart, a small manuscript by Henry Moriarty, a Royal Navy officer who had been appointed to navigate the *Great Eastern*, shows the position of cable and ship between 3 and 11 August as the crew tried in vain to recover the broken end. For over a week, Moriarty and those working under him calculated positions and navigated backwards and forwards over the cable. The crew managed to catch it three times with a grapnel – a specially barbed anchor – but were unable to bring it to the surface. This sketch shows the final attempt: where they were at noon on 11 August when they lowered the grapnel; where they were at 3.35pm when they hooked the cable; where they were at 6.50pm when the rope broke, leaving them without enough to try again. The following year, 1866, a Transatlantic cable was successfully laid. Using the locations recorded in 1865, the broken end was also raised and spliced to make a second working connection.

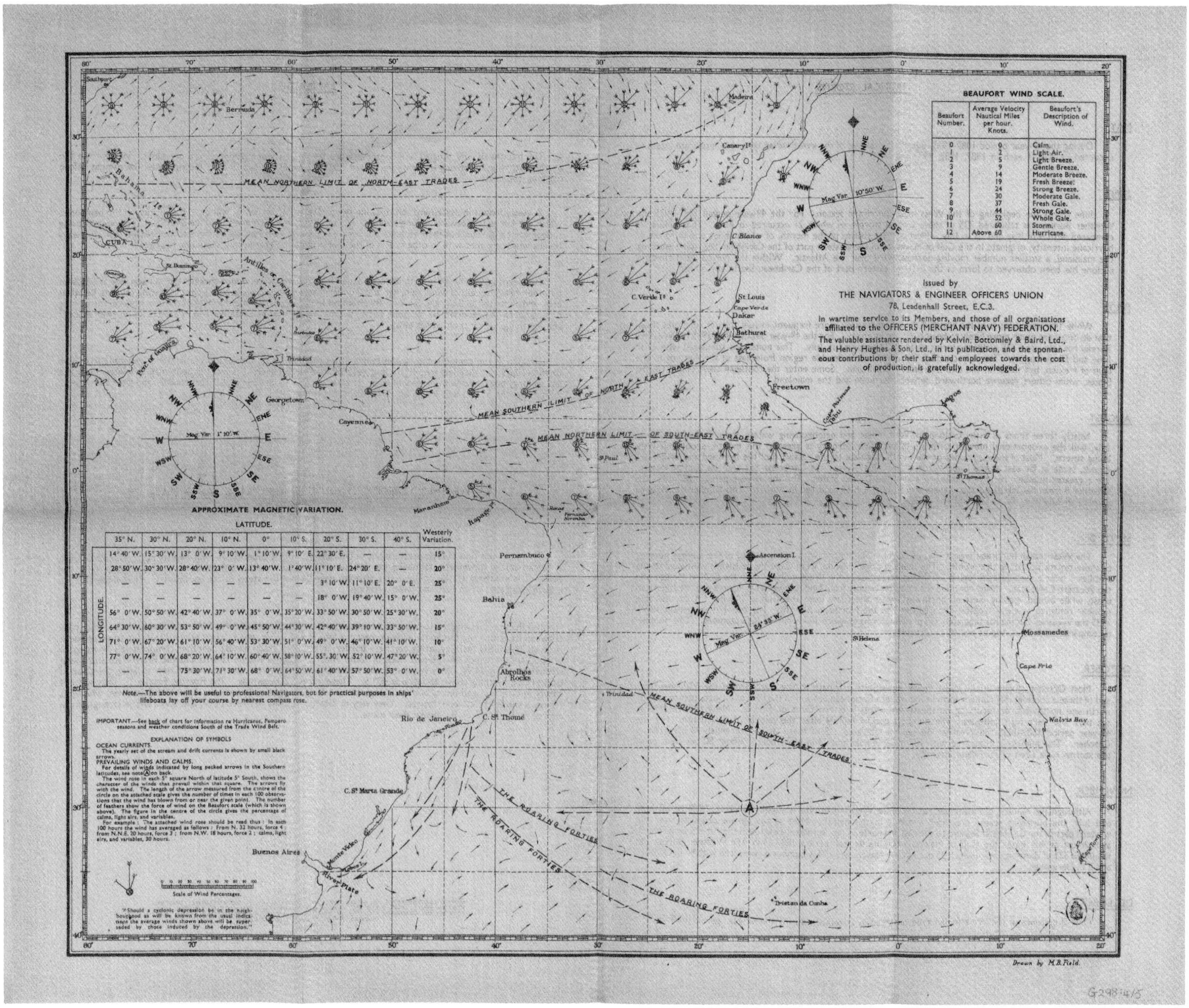

*Chart of the South Atlantic Ocean specially prepared for and issued by the Navigators & Engineer Officers Union*, Navigators and Engineer Officers Union, 1942

During the Second World War, which saw the sinking of thousands of merchant vessels, provision for seafarers forced into lifeboats became a pressing issue. Alongside the need for airtight first-aid kits and nonperishable food went discussion of emergency navigational equipment. In Britain, emergency charts were first supplied by the Navigators and Engineer Officers Union, formed in 1935 to push for improved working conditions and better pay for officers in the Merchant Navy. In the early 1940s, as well as arguing for official supply of better equipment for lifeboats, the Union itself produced a chart of the North Atlantic and then the one of the South Atlantic, pictured here. Intended to be kept in the pocket in case of emergency need while at sea, the charts were printed on waterproof fabric and in waterproof ink. Showing the prevailing winds and currents in different parts of the Atlantic,

they were intended to help people in lifeboats set a course for land. In September 1942 the Union's magazine, *The Navigator*, published a letter in which a member described 'how very useful we found the Union's Boat Chart when I was recently torpedoed … Believe me, all hands in the ship's lifeboat will never forget the real service which you rendered to us by issuing these charts.'[33]

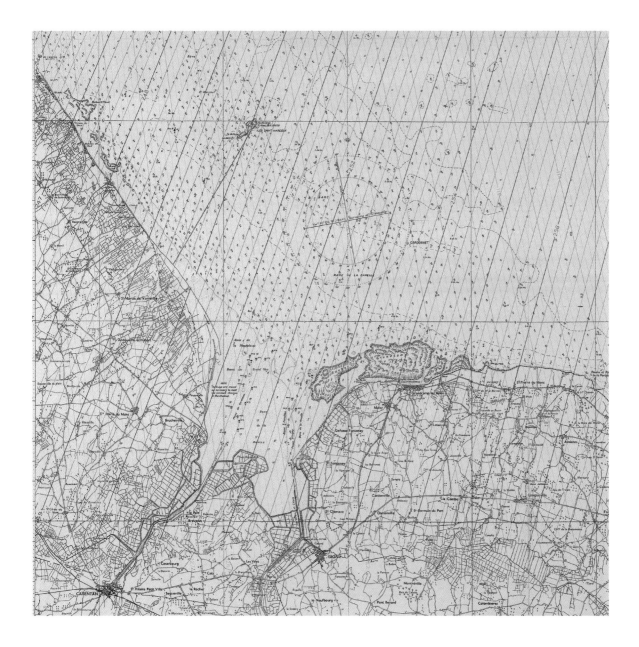

'To me it was pure magic.' That was how, looking back at the D-Day Landings, Rear-Admiral Steve Ritchie described the radio navigation system that was key to the success of landing Allied forces on the coast of Nazi-occupied France in 1944.[34] The magic was created by gramophone and record company, Decca, whose expertise in radio research facilitated the development of a top-secret and extremely precise system of coastal navigation. Through a combination of radio signals, measured from fixed coastal beacons by receivers on a vessel, and the corresponding lattice of criss-crossing curved lines printed on this chart, navigators were able to establish their position. When news of the navigation system was made public in 1945, it was claimed to be five times more accurate than any other navigation system.[35] Decca established a separate company for developing and producing this system, and it went on to become hugely successful. By 1970 it had 30,000 marine users, though towards the end of the twentieth century it gradually fell out of use as satellite technologies such as GPS (Global Positioning System) became more widely available. The last Decca transmitters in England were switched off on 31 March 2000.[36]

*Iles St Marcouf to Cap Manvieux F.1015*
(DECCA), Hydrographic Department, 1944

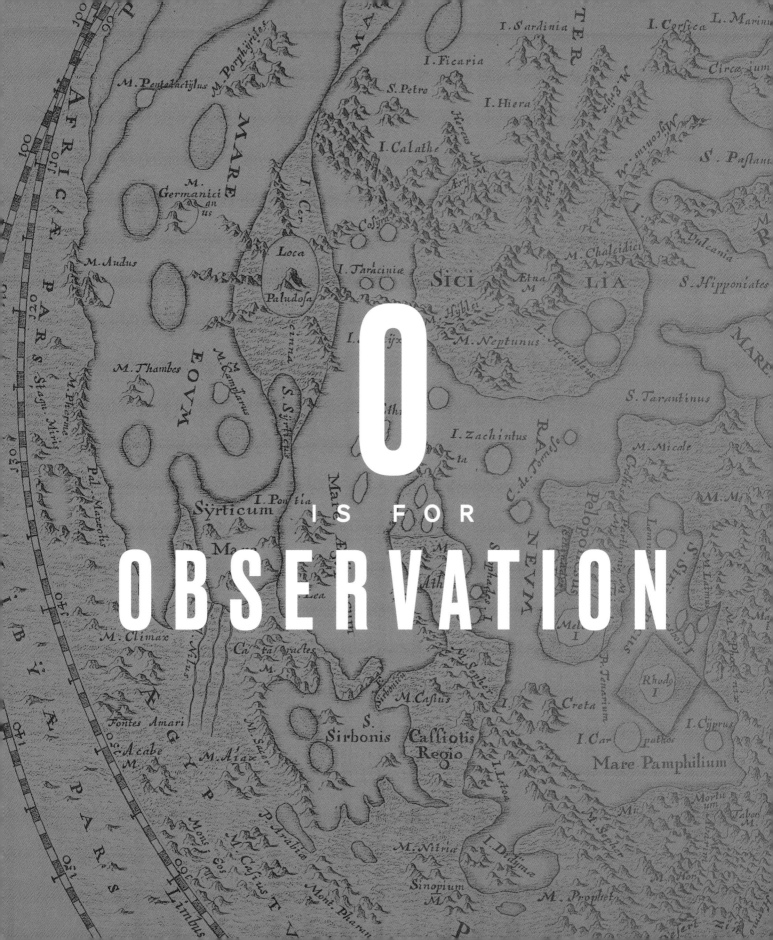

O

IS FOR

OBSERVATION

Whether observing the weather with a barometer, the Moon with a telescope, or the stars with the naked eye, how, when, where and why observations are made and communicated are some of the key questions in the history of science. Such questions reveal observations as historically contingent, possible, desirable and comprehensible because of specific circumstances. Whether investigations of the surface of the Moon, the geography of the stars, or of conditions at sea, maps used to communicate observations are shaped by concerns about what will appear trustworthy to their intended audience and what is economically or politically expedient.

European concepts of global geography shifted substantially during the sixteenth century, largely as a result of long-distance imperial and trading voyages. But it was not just the understanding of terrestrial geography that changed. For centuries, European and Islamic astronomers had been very familiar with the constellations of the northern hemisphere but knew much less about the geography of the stars around the South Pole. In the 1590s, the newly independent Dutch Republic sent an expedition to East Asia to assess possible routes there and the prospects for establishing a trading colony. Mapping the southern stars became part of the imperial venture. The navigators Pieter Dirkszoon Keyser and Frederick de Houtman were instructed to map them by Petrus Plancius, a prominent Dutch astronomer and cartographer, who suggested the route for the voyage. The 135 stars they observed were later plotted into 12 constellations by Plancius, who used European ideas of an exotic south for their forms: toucan, chameleon, flying fish. These figures were first made public on celestial globes, such as this one by cartographer Jodocus Hondius, made in 1600. Heralded as a great national and scientific success for the new republic, the constellations derived from Keyser and Houtman's observations were entirely bound up with the beginnings of the Dutch empire.

Celestial table globe,
Jodocus Hondius, 1600

'Tabula Selenographia',
*Selenographia: sive, lunae description*, Johannes Hevelius, 1647

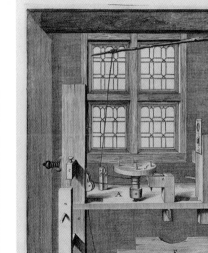

In the early seventeenth century, the invention of the telescope allowed astronomers to look at celestial bodies in more detail than ever before possible. Large and close, with a variety of textures on its surface, the Moon was an obvious object for their curiosity. This new way of looking was not straightforward, though, and certainly not easy to share. One eyepiece meant only one observer, which in turn meant that sharing the results of observations so that they might appear trustworthy and the astronomer might be respected as an authority was an ongoing challenge. In 1647 Johannes Hevelius (1611–87) produced a book detailing his telescopic study of the Moon, the *Selenographia* (1647). He devoted the first half of the work to showing readers not what he had seen, but how he had seen it: how he ground the lenses; how he built his telescope (above, right). He pictured himself, too, hand on hip at his telescope, showing the stability of its stand (above, left). In common with most accounts of natural philosophy in the seventeenth century, this intensive description served to bring readers as close as possible to the actual observations, making them virtual participants in the results, such as in this map of the Moon.

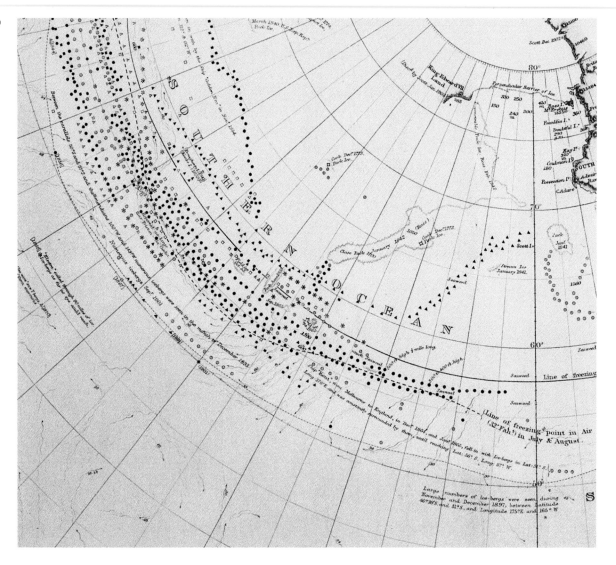

This chart brings together some 130 years of ice observations, with symbols representing icebergs seen in different months of the year and annotated textured lines showing the extent of pack ice. It was made because of radical changes to routes travelled at sea, which resulted from the development of the steam-ship. Navigators of steam-ships, newly able to take the shortest geographical route across the ocean, rather than relying on patterns of winds and currents, chose to go closer and closer to Antarctica on journeys to Australia. As they did so, the likelihood of being impeded or imperilled by the ice increased. Using historic observations as a data set, from the expedition of Captain James Cook in 1772–75 to that

of Ernest Shackleton 1908–09, the chart also marks those latitudes likely to be 'more doubtful and indeed dangerous', especially in July and August (the Antarctic winter), while also cautioning sailors that an absence of ice in previous years does not imply an area safe from danger. Where today we are more likely to see historic observations of marine environments used to understand the impacts of climate change, this ice-chart was used to facilitate the voyages of coal-powered steam ships.

*Ice chart of the southern hemisphere,*
**Hydrographic Office, 1910 (first published 1874)**

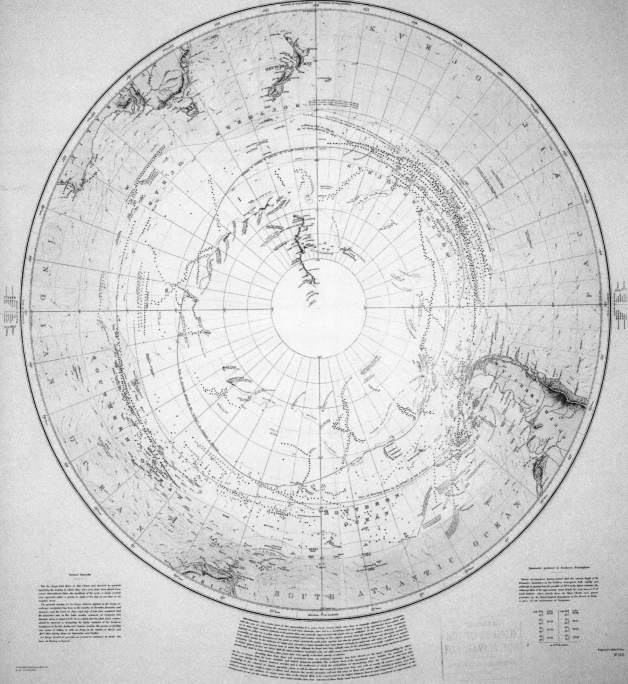

# ICE CHART

## OF THE SOUTHERN HEMISPHERE

Compiled from the Voyages of

Cook 1772-5, Bellinghausen 1820-21, Weddell 1822-4, Foster 1828-9, Biscoe 1830-2, Balleny 1839, D'Urville 1838, Wilkes 1839, Ross 1840-2-3, Scott 1901-4, and Shackleton 1908-9.

These Ice-papers on Ice-bergs in the Southern Ocean by M[r] Towson 1859-60, from the 19th Meteorological papers (based of Tracks) Nautical documents in Hydrographic Office.

1874

SOUNDINGS in FATHOMS.

London, Published at the Admiralty 10th Aug[t] 1874, under the Superintendence of Rear Admiral G.H. Richards, C.B.F.R.S. Hydrographer.

## OCEAN AREAS WHERE THE NUMBER OF METEOROLOGICAL OBSERVATIONS IS INADEQUATE ZONES OCEANIQUES OU LE N

All ships under way in ocean areas coloured in blue are requested to send radio weather reports to the coastal radio stations shown on the map.

Tous les navires faisant route dans les zones oceaniques o

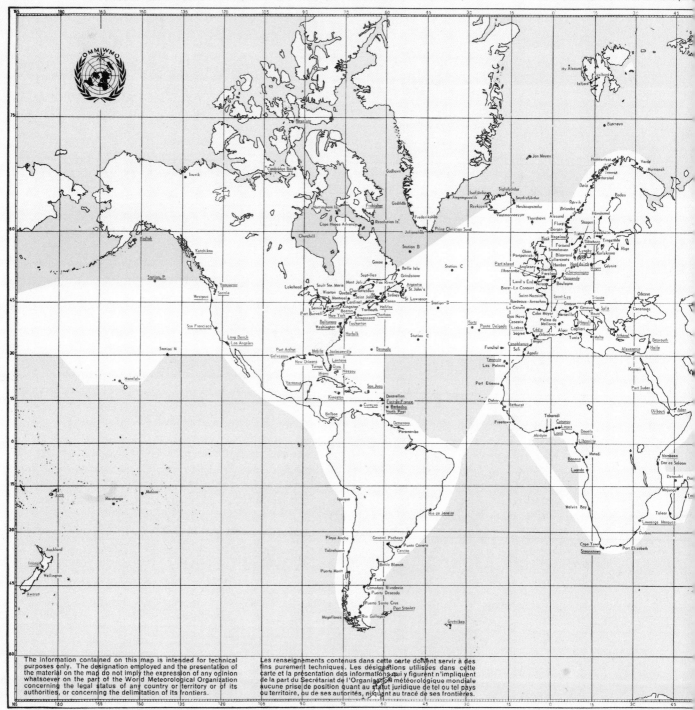

The information contained on this map is intended for technical purposes only. The designation employed and the presentation of the material on the map do not imply the expression of any opinion whatsoever on the part of the World Meteorological Organization concerning the legal status of any country or territory or of its authorities, or concerning the delimitation of its frontiers.

Les renseignements contenus dans cette carte doivent servir à des fins purement techniques. Les désignations utilisées dans cette carte et la présentation des informations qui y figurent n'impliquent de la part du Secrétariat de l'Organisation météorologique mondiale aucune prise de position quant au statut juridique de tel ou tel pays ou territoire, ou de ses autorités, ni quant au tracé de ses frontières.

Ed. 1965. (1.VI.1966)

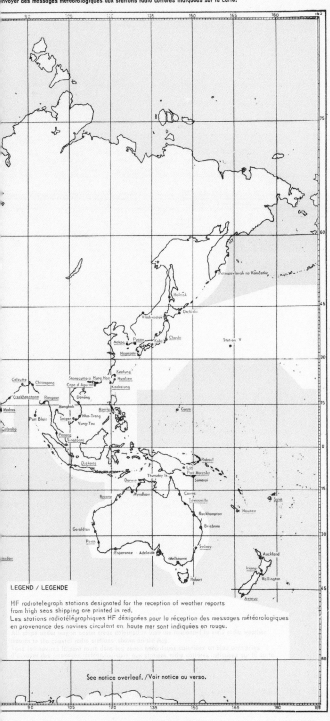

Weather is a global phenomenon, and understanding it involves worldwide effort. That was one of the key principles of the World Meteorological Organization (WMO), founded in the middle of the twentieth century to foster greater cooperation between national meteorological offices and to develop global systems of data collection and distribution. It supported newly decolonised states in developing their own meteorological services and facilitated international exchange of weather data even at the very height of the Cold War. Voluntary Observing Ships formed a key network of observers. These were merchant vessels supplied with instruments whose crews were trained by the WMO to make weather observations – of atmospheric pressure, wind direction and speed, and sea surface temperature – and transmit them by radio. This facilitated marine weather forecasting, a life-saving service broadcast to shipping using radio technology. But there were vast areas of ocean where most ships, sticking to the fastest routes between ports, did not travel. This map shows ocean areas where weather observations were so very sparse that building forecasts was not possible. Made by the WMO, it was part of an attempt to encourage vessels passing through such waters to make and submit weather observations, describing any data from these regions as of 'extremely great value'.[37]

*Ocean areas where the number of meteorological observations is inadequate*, World Meteorological Organization, 1965

P

IS FOR

PAPER

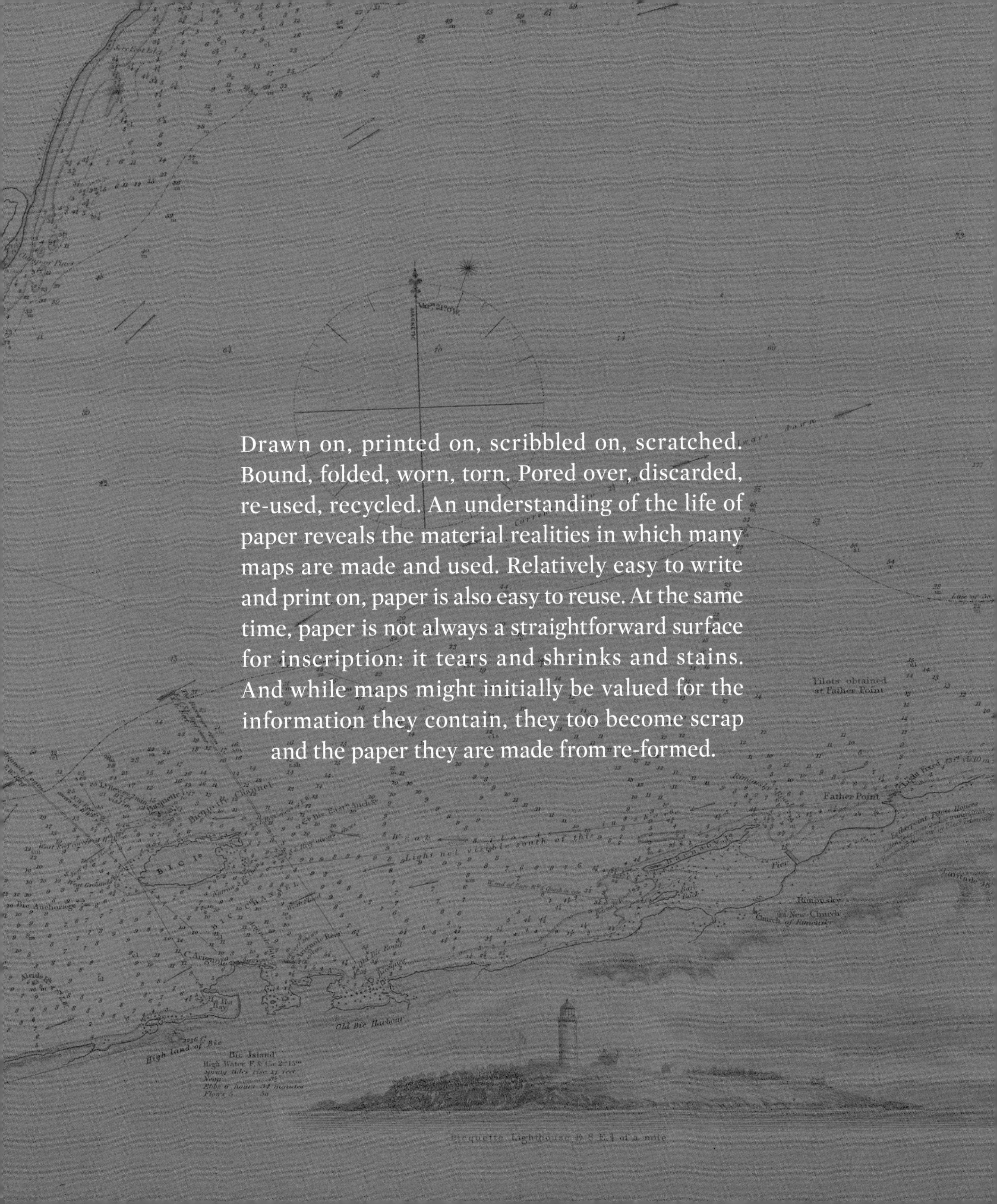

Drawn on, printed on, scribbled on, scratched. Bound, folded, worn, torn. Pored over, discarded, re-used, recycled. An understanding of the life of paper reveals the material realities in which many maps are made and used. Relatively easy to write and print on, paper is also easy to reuse. At the same time, paper is not always a straightforward surface for inscription: it tears and shrinks and stains. And while maps might initially be valued for the information they contain, they too become scrap and the paper they are made from re-formed.

For surveyors, engravers and printers, paper was not just a surface worked on, but a material worked with. How far this was the case became apparent in a dispute that arose between the Hydrographic Office in London, which published charts for the Admiralty, and Henry Wolsey Bayfield (1795–1885), a Royal Naval surveyor employed on the St Lawrence River, a vast waterway which was the main route into the British colony of Lower Canada. Bayfield, who sent manuscript surveys back to London to be engraved and printed, noticed inconsistencies in measurements in his published work that led him to allege carelessness on the part of the engravers. To those working in highly skilled roles for the Hydrographic Office in London this was quite an affront. Such inconsistencies, they informed Bayfield, were typical simply because paper, which was dampened before printing, warped and shrank irregularly as it dried. It was found that a printed sheet of paper measuring 25 by 40 inches would shrink by 0.2 to 0.6 of an inch. That does not sound like much, perhaps, but on Bayfield's St Lawrence plans, it could be the equivalent of up to 0.6 of a mile. And in a profession based on the setting down of precise measurements on paper, the warping and shrinking of that very medium was a cause for serious, but ultimately unresolvable, concern.

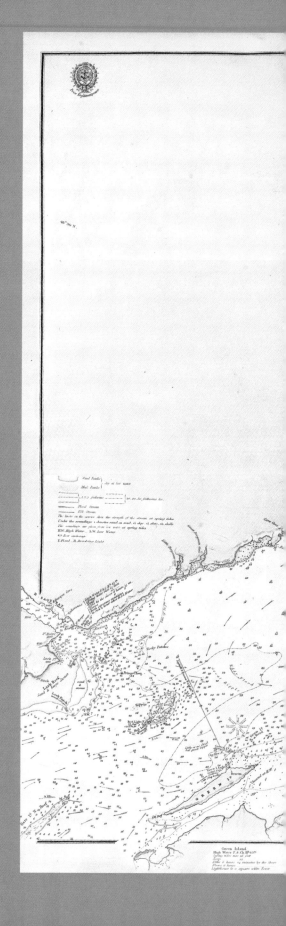

*Plans of the River St Lawrence below Quebec:*
*Sheet 2*, Henry Wolsey Bayfield and the
Hydrographic Office, 1865 (first published 1837)

PLANS OF THE

# RIVER St LAWRENCE

*BELOW QUEBEC*

## Sheet 2

BETWEEN THE RIVERS

**BERSIMIS** AND **SAGUENAY**

INCLUDING

**BIC** AND **GREEN ISLANDS**

SURVEYED BY CAPT. BAYFIELD R.N. F.A.S.

1827 _ 1834.

*Variation in 1832, increasing annually about 5 minutes*

SOUNDINGS IN FATHOMS

Most globes are damaged, at least a little bit. Made as papier-mâché hemispheres, covered in plaster and then in paper segments known as gores, before being mounted in the metal and wooden framework within which they rotate, there is ample opportunity for these fragile objects to break. The cracked shell of this globe, made in 1842 by London-based maker Newton and Son, means that we can look inside. A globe-maker would often place sheets of paper between the hemisphere and the mould on which it was shaped to act as a 'release', allowing the two to be easily separated. The cheapest source of available paper was scrap – often proof pages from publishers, printed for final typesetting checks – and the most widely published book of the time was the Bible. Little wonder, then, that inside this globe we find pages from various books of the Bible (opposite, bottom). Only the surface of a globe was ever intended to be visible. What sits inside tells us something about practices of economy and re-use in the making of these complex objects.

Terrestrial table globe,
Newton & Son, 1842

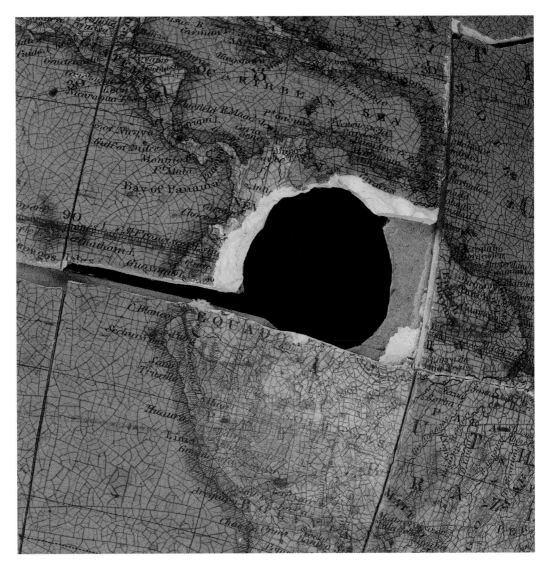

Boat drill. March 23

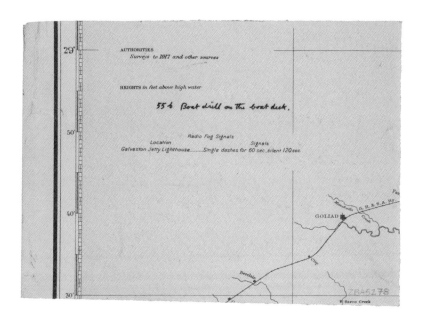

Fragments of US Coastal and
Geodetic Survey Chart 'United
States – Gulf Cost – Galveston
to Rio Grande'/'Boat Drill' and
'Boat Drill on the Boat Deck',
John Kingsley Cook, 1941

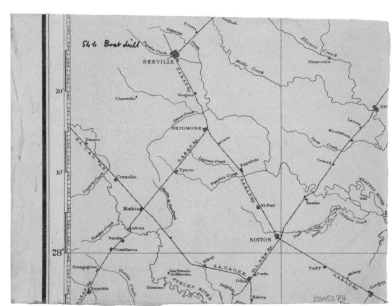

When John Kingsley Cook (1911–94) joined the Merchant Navy as
a wireless operator during the Second World War, he was already a
Royal Academy-trained artist. Although not an official war artist, he
recorded his experiences at sea and as a prisoner of war in Algeria in
both on-the-spot sketches and later recollections. He first crossed
the Atlantic in December 1940, a journey that he, like many merchant
seafarers, found arduous and nerve-wracking. He did, however, find
time to draw. These drawings and watercolours were made in spring
1941 and show different elements of a boat drill, essential practice
especially in the wartime Merchant Navy, when so many vessels
were sunk, forcing their crews to take to lifeboats. For his drawing
surface, Cook made use of the ship's obsolete charts. These images
were made on the back of a piece of US Coast and Geodetic Survey
chart No. 1117, 'United States – Gulf Coast, Galveston to Rio Grande'.
As key navigational aids, charts are regularly updated and reissued,
meaning they go out of date and need to be replaced frequently. Cook
himself noted how the paper 'can take pen and ink beautifully'.[38] In
the context of wartime paper rationing, introduced in February 1940,
the paper used for charts would have been some of the finest quality
on board.

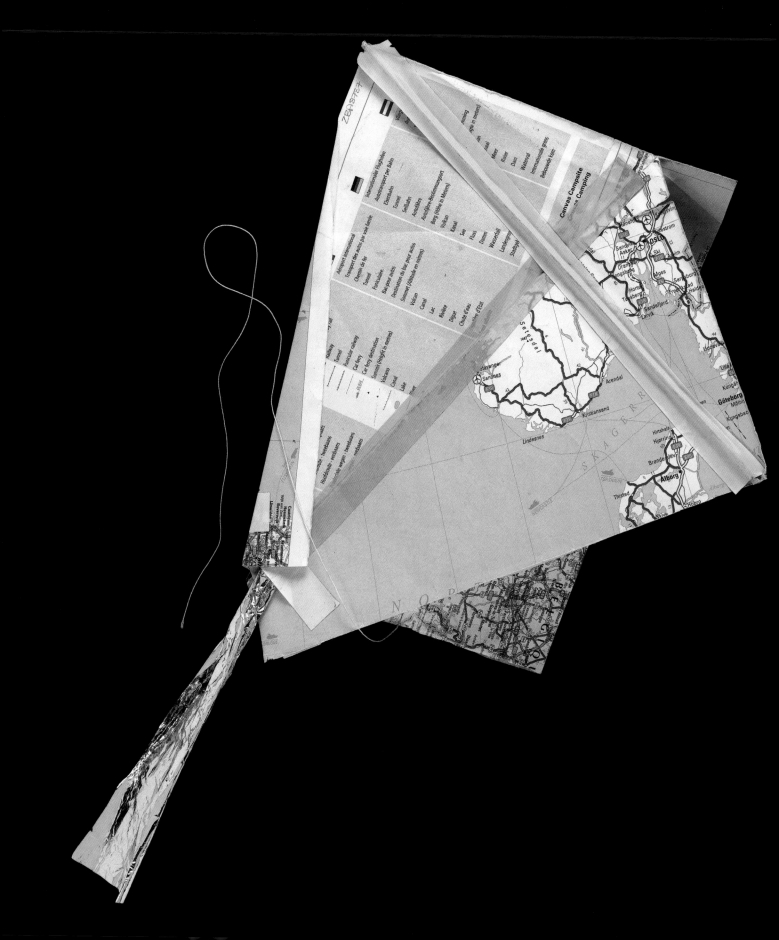

Kite made from a road map,
Art Refuge UK, 2015

Since 2015, art and art-therapy organisation Art Refuge UK has been working with refugees in Calais, northern France. Using the limited infrastructure of the camp known as 'the Jungle', which was demolished by the French authorities in 2016, the artists working with the charity started to make kites alongside Afghan and Kurdish residents who had already been doing just that. As an activity, kite-making provided opportunities for creativity, problem-solving and collaboration, in a context where people's traumatic experiences, both past and present, can lead to an acute sense of psychological unsafety. This kite, made by one of the Art Refuge UK workers, was crafted from a road map, some electrical tape and a piece of emergency blanket. This was both pragmatic and symbolic. Pragmatic, because road atlases provide large sheets of robust paper which can withstand the Calais wind. Symbolic, because, in a context where people's movement across borders is severely restricted, a flying map appropriates an object that is often the primary expression of those very borders. In the end, this kite did not fly very well. The paper was too heavy and the method of construction also made it less likely to catch the wind than those made by the Afghan and Kurdish makers. And perhaps, ultimately, nothing could bear such a symbolic load.

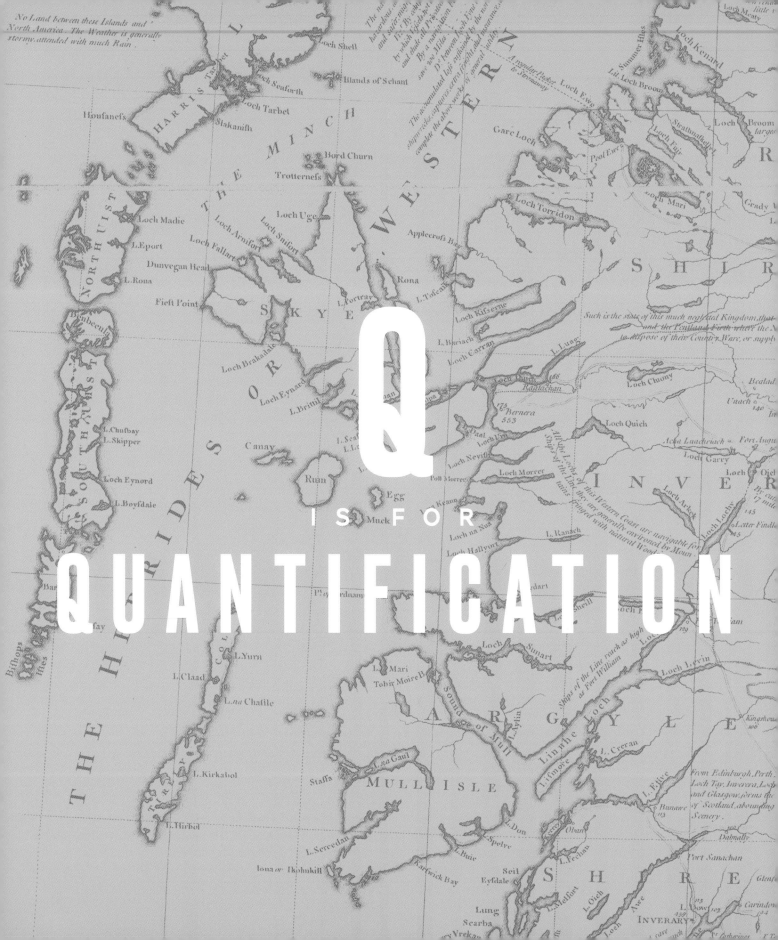

Q

IS FOR

QUANTIFICATION

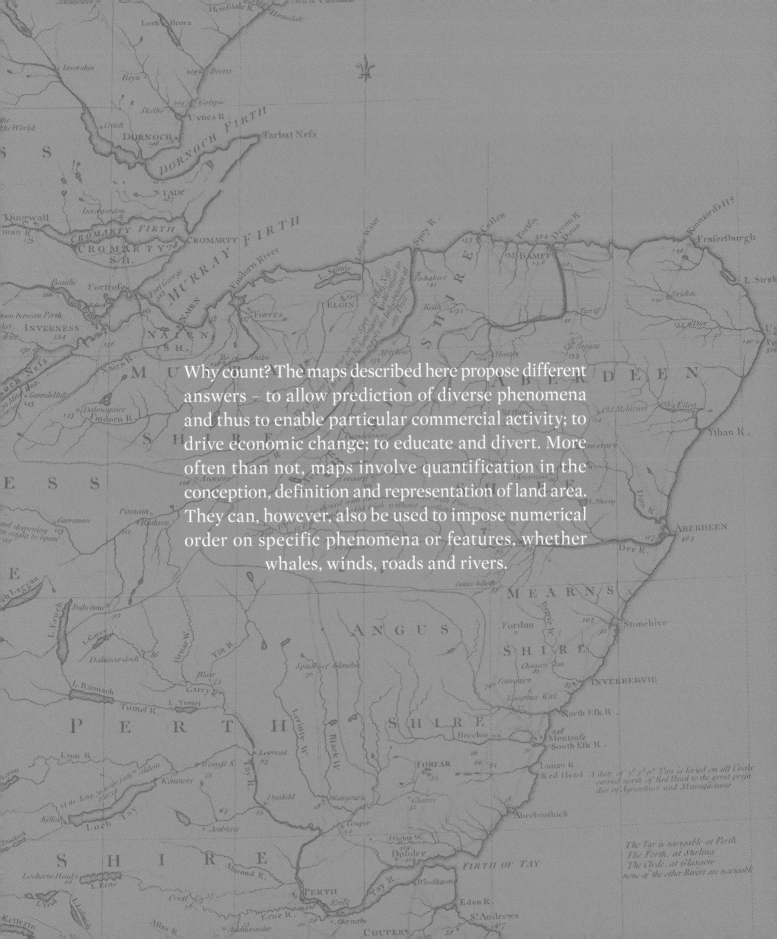

Why count? The maps described here propose different answers – to allow prediction of diverse phenomena and thus to enable particular commercial activity; to drive economic change; to educate and divert. More often than not, maps involve quantification in the conception, definition and representation of land area. They can, however, also be used to impose numerical order on specific phenomena or features, whether whales, winds, roads and rivers.

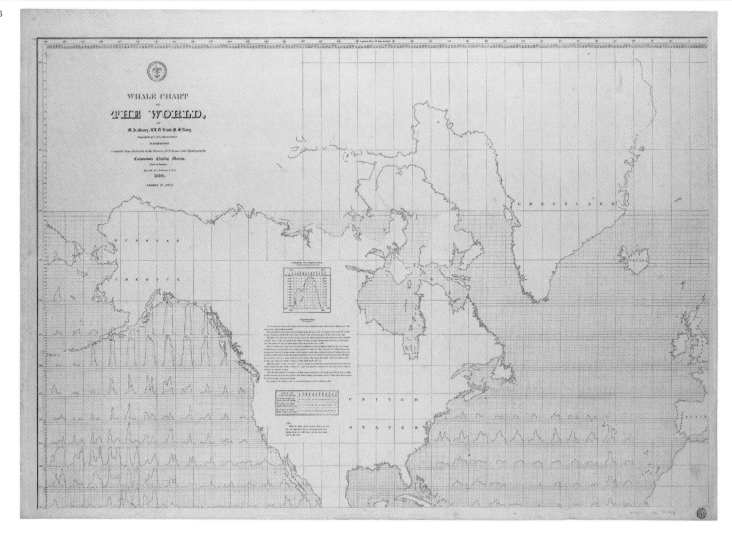

Towards the middle of the nineteenth century, concern with the distribution of whales motivated Matthew Fontaine Maury (1806–73), then Superintendent of the US Depot of Charts and Instruments, to compile and then plot data linked to them on a world map. Collecting and organising data about the presence of plants and animals to understand the regions they could be found in – known as biogeography – had become popular a few decades prior. Understanding the whereabouts of whales had a distinct commercial purpose, especially in the USA, where by 1850 whaling was the fifth largest industry. Valuable whale oil, as well as the even more valuable spermaceti (from the head of sperm whales), was used for lighting and to lubricate machinery. Ambergris, found in the sperm whale's intestines, was used to make perfume and many common household objects were made from whalebone. Maury's map brought together data collected by whaling vessels and, in five-degree squares divided for the months of the year, plotted the number of days vessels had been looking for whales in a particular area alongside the number of days that sperm whales and right whales were actually seen. Maury reflected that his work 'cannot fail to prove of great importance to the whaling interests of the country … which fishes up annually from the depths of the ocean, property, in real value which far exceeds that of the gold mines of California'. One captain who was sent copies of the chart described it as 'a precious jewel'.[39] Designed to further commercial activity, Maury's whale charts show how the study of ocean environments in the nineteenth century was not disinterested enquiry.

*Whale Chart of the World*, M.F. Maury,
**United States Hydrographical Office, 1852**

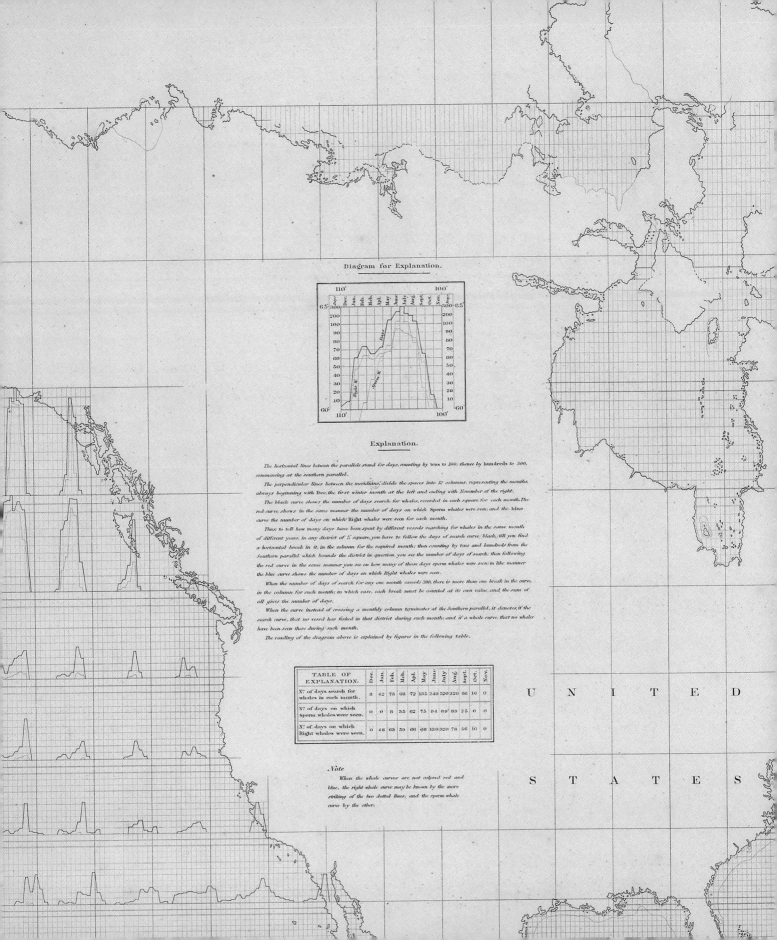

## Diagram for Explanation.

## Explanation.

The horizontal lines between the parallels stand for days, counting by tens to 100; thence by hundreds to 300, commencing at the southern parallel.

The perpendicular lines between the meridians, divide the spaces into 12 columns, representing the months, always beginning with Dec. the first winter month at the left and ending with November at the right.

The black curve shows the number of days search for whales, recorded in each square for each month. The red curve shows in the same manner the number of days on which Sperm whales were seen; and the blue curve the number of days on which Right whales were seen for each month.

Thus, to tell how many days have been spent by different vessels searching for whales in the same month of different years, in any district of 5° square, you have to follow the days of search curve, black, till you find a horizontal break in it, in the column for the required month; then counting by tens and hundreds from the Southern parallel which bounds the district in question, you see the number of days of search; then following the red curve in the same manner you see on how many of those days sperm whales were seen; in like manner the blue curve shows the number of days on which Right whales were seen.

When the number of days of search for any one month exceeds 300, there is more than one break in the curve, in the column for such month; in which case, each break must be counted at its own value, and the sum of all gives the number of days.

When the curve instead of crossing a monthly column terminates at the Southern parallel, it denotes, if the search curve, that no vessel has fished in that district during such month; and if a whale curve, that no whales have been seen there during such month.

The reading of the diagram above is explained by figures in the following table.

| TABLE OF EXPLANATION. | Dec. | Jan. | Feb. | Mch. | Apl. | May | June | July | Aug. | Sept. | Oct. | Nov. |
|---|---|---|---|---|---|---|---|---|---|---|---|---|
| Nº of days search for whales in each month. | 8 | 62 | 73 | 66 | 72 | 135 | 240 | 520 | 320 | 66 | 16 | 0 |
| Nº of days on which Sperm whales were seen. | 0 | 0 | 8 | 35 | 62 | 75 | 94 | 89 | 83 | 25 | 0 | 0 |
| Nº of days on which Right whales were seen. | 0 | 48 | 63 | 59 | 68 | 68 | 130 | 320 | 78 | 56 | 10 | 0 |

*Note*

When the whale curves are not colored red and blue, the right whale curve may be known by the more striking of the two dotted lines, and the sperm whale curve by the other.

U N I T E D

S T A T E S

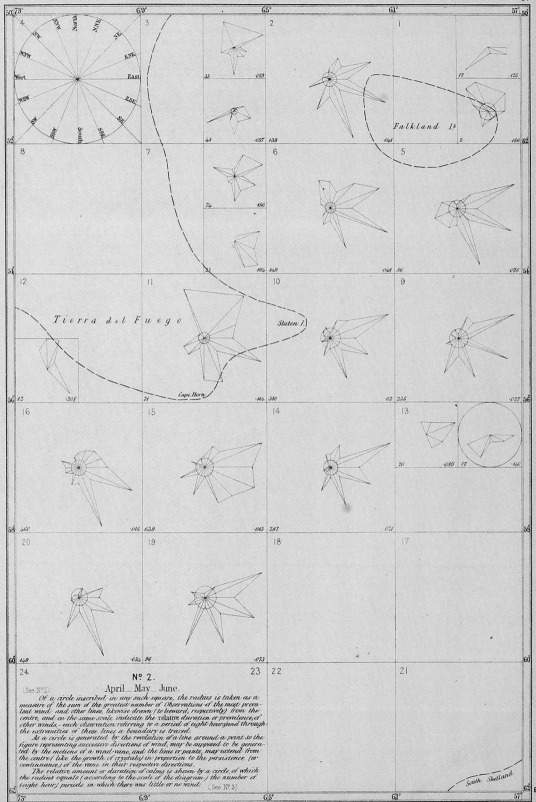

Falkland I.ˢ

Tierra del Fuego

Staten I.

Cape Horn

No 2.

April — May — June.

(See Nº 1.)

Of a circle inscribed in any such square, the radius is taken as a measure of the sum of the greatest number of Observations of the most prevalent wind, and other lines, likewise drawn (to leeward, respectively) from the centre, and on the same scale, indicate the relative duration or prevalence, of other winds, — each observation referring to a period of eight hours, and through the extremities of these lines a boundary is traced.

As a circle is generated by the revolution of a line around a point so the figure representing successive directions of wind, may be supposed to be generated by the motions of a wind-vane, and the lines or points, may extend from the centre (like the growth of crystals) in proportion to the persistence (or continuance) of the vane in their respective directions.

The relative amount or duration of calms is shewn by a circle, of which the radius equals (according to the scale of the diagram) the number of (eight hour) periods in which there was little or no wind. (See Nº 3)

South Shetland

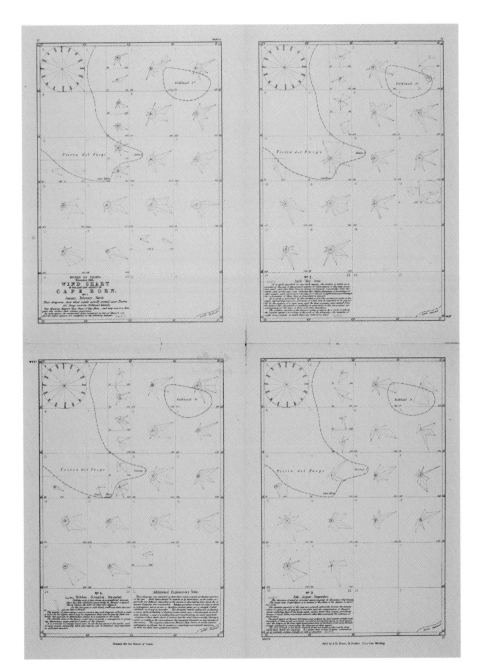

*Wind chart, showing winds for the four quarters of the year – Cape Horn – Eastern Vicinity*, Meteorological Department of the Board of Trade, 1855

Making complex data accessible at a glance was a very Victorian undertaking. This map, illustrating the prevailing winds around Cape Horn, was intended to help navigators plot their courses more efficiently. The wind charts were some of the first publications issued by the British Meteorological Department of the Board of Trade, which had been founded in 1854 to support commercial shipping through the study of the weather. There are four charts per region, which each represent a quarter of the year. The ocean is divided into squares, within which is an irregular polygon whose points relate to the points of the compass. The more frequently the wind had been recorded blowing in a particular direction, the further out from the centre of the polygon a point was drawn. Straight lines then connected them, forming what Robert FitzRoy (1805–65), head of the Meteorological Department, named 'wind stars'. There is also a circle to indicate the number of calm days, when there was little or no wind – the larger it is, the more calm days were observed. Here they are all small, sometimes practically invisible, because of the ferocious conditions around Cape Horn. As astronomer Charles Piazzi Smyth (1819–1900) noted of the wind stars, 'a moment's glance ... may give instant and true judgement'. They allowed observations from multiple voyages, which provided quantities of numerical data too overwhelming to take in, to be represented more simply.[40]

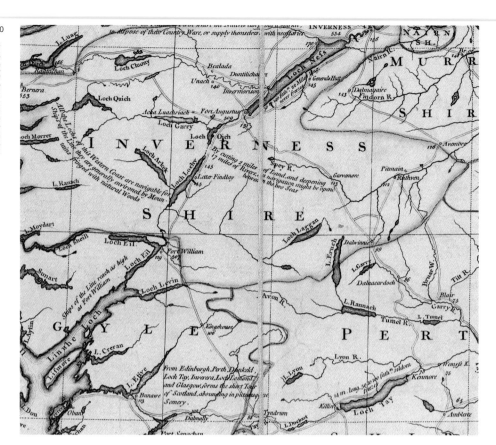

Presenting the Scottish Highlands as an untapped natural resource, this map by Scottish-born London bookseller John Knox (1720–90) was a product of the so-called 'Improvement movement' in eighteenth-century Britain. Knox was particularly interested in the Scottish Highlands and the region's economic transformation. The Highlands suffered social catastrophe in the eighteenth and nineteenth centuries, as people were forced off land they had lived on and worked for generations as part of the clan system to be replaced by large, profitable sheep farms, leaving many destitute in the name of agricultural progress. Reformers like Knox used these developments to advocate for changes that would shift people away from regional subsistence economies towards commercial societies in the north of Scotland and enable the natural resources of the region to be better

exploited.[41] The Improvers believed this would also lead to greater social and political cohesion across the land, still a pressing issue in the decades following the Jacobite uprisings, in which many Highland clans had joined the attempt to restore James VII of Scotland and II of England and then his descendants to the throne. Knox primarily advocated infrastructural change, in particular building canals and establishing fisheries with planned villages to serve them, so that the natural resources of the region might be better exploited. For that, he needed numbers – distances along roads and between ports, the cost of digging canals – all of which are included on this map. Along with annotations detailing the abundance of herring and the inadequacy of communication routes, the map was one of the ways Knox argued for the transformation of Scotland.

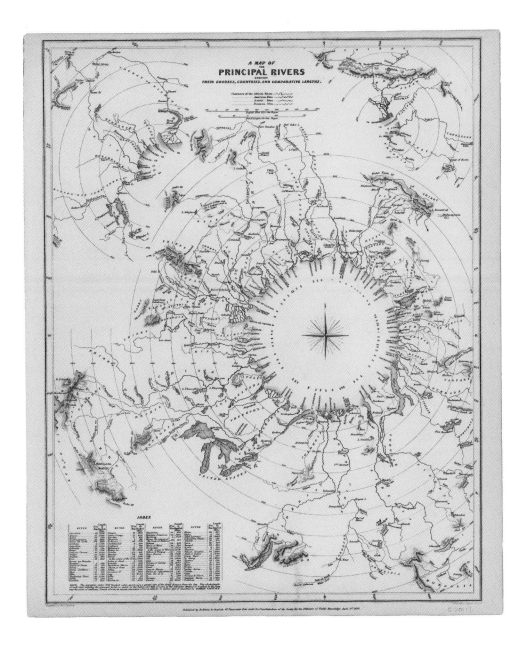

*A Map of the Principal Rivers,*
*shewing their courses, countries*
*and comparative lengths,*
Society for the Diffusion of
Useful Knowledge, 1834

With the Po next to the Amazon and the Tigris and Euphrates next to the Rio de la Plata, this map rearranges world geography with one purpose: to compare rivers. Arrayed around imaginary seas into which the different waters flow, their courses are plotted over a series of concentric circles, each representing 200 English miles. In one corner, the reader is invited to consider their differing lengths in detail: the Nile in Egypt at 2,750 miles; the Liffey in Ireland at 52. This was one of the more unusual maps published by the Society for the Diffusion of Useful Knowledge, which had been established in the turbulent 1820s by the great and the good of Regency England with a view to educating the working classes. The guiding idea was that 'the peace of the country and the stability of government, could not be more effectively secured' than by such endeavours.[42] In particular, the Society sought to provide

cheap, authoritative printed works, which included a remarkable series of around 200 maps, covering 'the various branches of Knowledge'. The work of the Society ensured that maps, carefully designed and printed from engraved copper, were available to less well-off readers for the first time, in an era when interest in world geography, and in British colonies, was growing. This document is one example of a nineteenth-century genre of comparative map, in which disparate geographical features are brought together for visual effect. Some have mountains jostling against one another; others feature the world's islands arranged side by side. At a time when geographical learning tended to be thought of as the amassing of facts – name, location, length – such a collapsing of place, isolating rivers from their local contexts, was a diverting way of making global comparisons.

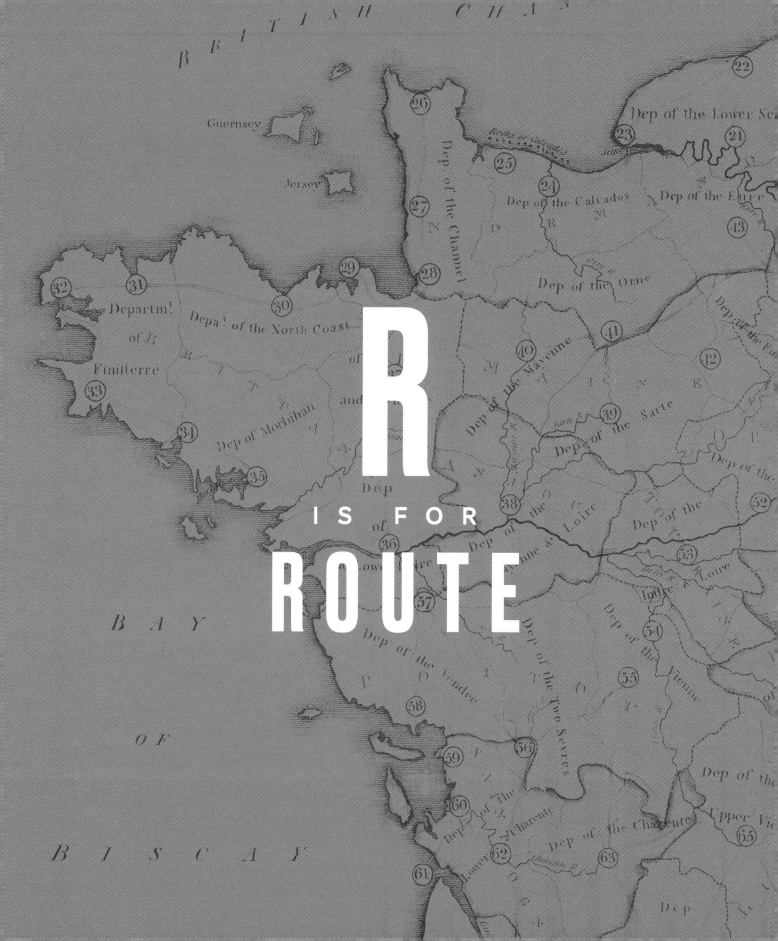

# R

## IS FOR

# ROUTE

What do we learn about a map from the route it shows? What do we learn about a route from the maps that show it? Guides to possible journeys, real or imagined; marks of routes taken, memorialising particular histories while ignoring others; plans for changing routes. Many maps either propose or celebrate particular ways of travelling. Route maps can make us think about available information, too, and what datasets are accessible or desirable in different circumstances, whether remote journeys in the eighteenth or precarious border crossings in the twenty-first century.

*A new geographical pastime exhibiting a complete tour thro France* J. Enouy, 1795

For the elite and middling classes of eighteenth-century Britain, touristic travel to the continent became an important form of extended leisure. Perceived to have educational and health benefits, a trip abroad also gave travellers a certain cachet at home. This game, a map of France with towns and cities marked by numbered circles, referring to descriptions in the margin, had players travel remotely from the comfort of their homes, using ivory pillars as counters to move across the paper sheet. The text provided them with an array of facts, which could, in turn, be inserted into polite conversation or spur reminiscence about past travel. It was, then, also a form of social instruction. The cathedral at Coutances in north-west France is 'one of the handsomest pieces of Gothic architecture'; Angouleme, in the south-west of the country, is famous for its glass manufactory'; the countryside around Nice

on the Mediterranean coast is 'delightful' ('stay one turn to see it'). Published in 1795, a decade during which Britain spent more time than not at war with France, travel for leisure was severely limited. The board also noted recent events: landing at Valenciennes, which had surrendered to the Duke of York in 1793, meant a player was allowed to skip ahead from stop 4 to 18; at Lyon, a massacre that same year would take them back 49 stops and cause them to miss two turns. The winner was the first to arrive in Paris – 'inhabitants reckoned at 1,000,000' – and earned a recommendation for a well-known guidebook to complete the tour at a distance in style.

*A new geographical pastime exhibiting a complete tour thro France* J. Enouy, 1795

3. CAMBRAY. Upon the river Scheld, large and handsome City; from whence came all the Chimney Sweepers of Flanders, called, jocosely, "The Cupids of Cambray."

4. VALENCIENNES. Upon the Scheld, taken from the Spaniards by Louis XIV. in 1677. In consequence of the surrender of this place to His Royal Highness the Duke of York, in 1793, the traveller will be conducted to Amiens.—No. 18.

5. MONS. Upon the Trouille, strong fortified Town; to this place, Monsieur, the King's Brother, escaped in his flight from France.

6. BRUXELLS. Upon the river Senne, capital of the Austrian Netherlands, large and handsome City. Stop three turns to see the curiosities of this place, viz. the Cathedral of St. Gudule, the handsome Tapestries, and the Cabinet of Natural History of Prince Charles.

7. ANTWERP. Upon the river Scheld, which is 675 feet wide, formerly the first City of trade in the Low Countries.

8. GHENT. Large fortified Town upon the Scheld, about 40 miles from Antwerp.

9. OSTEND. Free and commodious Port, well fortified; stop one turn to see the extensive and beautiful Quay, near one mile and a half in length.

10. LISLE. Upon the river Deule, capital of French Flanders, large and handsome City; stay one turn to see the citadel by Vauban, reckoned one of the strongest in Europe; its population is reckoned at 60,000

11. ARRAS. Upon the Scarpe; capital of Artois. In the centre of the Town is the celebrated Abbey of St. Vaast, founded by King Thiery, at the end of the 7th century; strong Citadel by Vauban.

12. ST. OMER. Upon the river Aa, tolerable large and well-built Town, having a College for English education; stay one turn to ride to Ardres, near this place, to see the Champ de Drap d'Or, where Henry VIII. and Francis I. had an interview.

13. DUNKIRK. Originally a Church built on the Sands, from which it takes its name, in the year 960. It belonged to the English in 1662, when the French purchased it with some other Villages, for the sum of five millions.

14. CALAIS. Tolerable well-built Town, containing near 10,000 Inhabitants. The City and Arsenal built by Cardinal Richelieu. Taken by famine in 1347, by Edward III. after 11 months siege.

15. BOULOGNE. Small sea-port at the mouth of the river Liane, two leagues North is Ambleteuse, where James II. landed, in 1688.

16. MONTREUIL. Pretty town on the South bank of the Canche, formerly near the Sea, now near four Leagues from it.

17. ABBEVILLE. Pleasantly situated on the river Somme, the birth-place of the Geographers Sansons, Duval, and Briet.

18. AMIENS. Capital of Picardy, upon the Somme, large and well-built City, containing 40,000 Inhabitants; stay one turn to see the Cathedral, which is one of the handsomest in France.

19. BEAUVAIS. Famous for its Tapestry, situated on the river Terrein.

20. PONTOISE. On the Oise. The Parliament of Paris was exiled here in 1720 and 1753.

21. ROUEN. Capital of Upper Normandy, large City, vessels of burthen come up to its Quays by the tide. Stop two turns to see the famous bridge of Boats, the Alley of St. Jean, Place d'Armes, and the place where the Maid of Orleans was burnt by the English in 1431.

22. DIEPPE. Tolerable sea-port Town, nearly destroyed by the English in 1694, since rebuilt as at present. The passage across the Channel to Brighton.

23. HAVRE. Commercial town, at the mouth of the Seine, it was bombarded by the English fleet, in 1759.

24. CAEN. Capital of Lower Normandy, on the junction of the Orne and Odon, well-built city, containing 40,000 inhabitants. In digging a canal from this town to the sea, in 1783, at the depth of 14 feet, was found some coins of Antoninus.

25. BAYEUX. Tolerable large Town, containing 14 parishes, situated on the river Aure.

26. CHERBOURG. Here you must stay two turns, to see the artificial Harbour, made by means of Cones, and loose Stones thrown between them, capable of containing a fleet of 80 sail.

27. CONSTANCES. The Cathedral is one of the handsomest pieces of Gothic architecture.

28. AVRANCHES. Tolerable large Town, from hence you must take a ride to Tinchebray, and stay one turn. Henry I. in 1106, gained a victory over his brother Robert, Duke of Normandy.

29. ST. MALO. A good Port, but difficult of access; strong and commercial Town. The birth-place of James Carter, the discoverer of Canada.

30. ST. BRIEUX. Tolerable town of Brittany, near the Sea; it has nothing remarkable.

31. MORLAIX. Small commercial Town, having a handsome Quay, and excellent manufactory for Snuff.

32. BREST. The first maritime department of France, and one of the finest Arsenals in the world. The port of Brest contains several fine Quays, surrounded by Magazines, filled with every necessary article for the Marine; stop two turns to view them.

33. QUIMPER. From hence you must go 11 leagues to Carhaix, where

66. CHATEAUROUX. Pretty town, on the river Indre. Here is a Castle and a considerable manufactory of Cloth.

67. GUERET. Upon the river Creuse, capital of the province of Marche.

68. UZERCHES. Situated on a Rock, at the foot of which runs the river Dordogne.

69. CLERMONT. Capital of Avergne, containing about 30,000 Inhabitants; stay one turn to taste its mineral waters, and see the Country round it, being covered with Vineyards and Meadows.

70. AURILLAC. Upon the little river Jordane, the most commercial Town of Avergne.

71. RHODEZ. Ancient and handsome City upon the river Aveiron, in the province of Guienne.

72. CAHORS. A city of Guienne, situated on the river Lot.

73. MONTAUBON. A city of Guienne, situated on the river Tarn, containing 11,000 Inhabitants, surrounded with country seats.

74. AGEN. Very pretty city, on the river Garonne.

75. LA REOLE. A Town on the river Garonne, between Bourdeaux and Agen; stay one turn to eat a poulets a la Crapaudine.

76. BORDEAUX. Large and commercial City, on the river Garonne; capital of Guienne; the entrance of its port defended by the Chateau Trompete; stay four turns to see the Theatre, the Chartron, the Cathedral, the Alley of Tourny, &c.

77. BAYONNE. Strong city, of great traffic, on the mouth of the river Adour, defended by two castles; stay one turn to taste its Hams.

78. BAREGE. Small town on a branch of the river Gave, celebrated for its mineral waters.

79. TARBES. A city of Gascony, on the river Adour.

80. AUCH. A city of Gascony, on the river Gers.

81. TOULOUSE. Large and handsome city, capital of Languedoc, on the river Garonne, containing 70,000 inhabitants; stay two turns to see the famous canal of Languedoc, which unites the Atlantic with the Mediterranean, about 64 leagues long, and 36 feet broad, 114 locks and sluices, began in 1666, and finished 1681, by Engineer Riquet, at the expence of 13 million of livres, or about 540,000l.

82. CASTRES. A town of Languedoc, nothing remarkable.

83. CASTELNAUDARY. A town of Languedoc, on an eminence.

84. MIREPOIX. Pretty city of Languedoc, on the river Lers.

85. CARCASSONNE. Ancient and considerable town, on the river Aude, famous for its manufactures of Cloth.

86. PERPIGNAN. Strong city, capital of Roussillon, having a Citadel, a University, public seminary, 12 Colleges, &c.

87. NARBONNF. Ancient and large city, about two leagues from the Sea, on the river Aude.

88. MONTPELLIER. Handsome city, about two leagues from the sea, having a university, an academy of arts and sciences; stay one turn.

89. LUNEL. A town of Lower Languedoc.

90. MENDE. An ancient town of Languedoc, on the Loire.

91. NISMES. Large and strong city of Languedoc; stay one turn to see the Amphitheatre, Fountain, &c. built 590 years before Rome.

92. ORANGE. Capital of a principality of the same name, governed formerly by Princes of its own (the House of Nassau) yielded to France by Treaty of Ryswick, 1697.

93. AVIGNON. A large city of Provence, on the East side of the Rhone, seven Popes resided in this City successively.

94. AIX. Capital of Provence, large and handsome City, containing 23,000 Inhabitants.

95. MARSEILLES. Large and commercial city; stay one turn to see the Academy of arts, the Citadel, and the superb Port, where were kept the King's galleys.

96. TOULON. Strong and famous port on the Mediterranean: its harbour consists of two large Basons: the new one contains the men of War and Frigates, and the old one the merchants' ships, having magazines for each Ship of that department.

97. FREJUS. Ancient city of Provence, near the Sea.

98. ANTIBES. Sea port of Provence, defended by a good Citadel and a square Fort, surrounded by handsome Walls.

99. NICE. A sea port on the frontiers of Italy; stay one turn to taste its Wine, and view the country round, which is delightful.

100. DIGNE. A town of Provence, nothing remarkable.

101. BRIANCON. A town of Dauphiny, on a branch of the river Durance, having a strong castle built on the front of a rock.

102. GRENOBLE. Handsome city, capital of Dauphiny, on the river Isere, having a cathedral, collegiate church, general hospital, &c.

103. VALENCE. Ancient city of Dauphiny, on the Rhone, between Lyons and Avignon.

104. LE PUY. A considerable city of Languedoc, near the source of the Loire.

105. LYON. Ancient, large, and handsome city, capital of Lyonois, on the conflux of the Saone and Rhone, having an academy of sciences and belles lettres, fine arts, public library, &c. containing 115,000 inhabitants, before the dreadful massacre in 1793. In consequence, the

The route drawn on this globe tells us who published it. In a way, that's quite surprising: the voyage track in question is that of Commodore George Anson (1697–1762), whose 1740–44 circumnavigation of the globe was very big news in eighteenth-century Britain. It was, mostly, a disaster. Around 1,300 of the almost 1,700 people who set out with the squadron died; of the eight ships that departed only one returned. But the stories of shipwreck and mutiny that emerged, and the fortune in Spanish silver with which Anson returned, made for accounts which were voraciously, and widely, consumed. In this period, any English map-maker working on a globe or a world map, would likely include Anson's most talked about route on their new product. So how does the route reveal anything about who made this globe? It is marked 'A New and Correct Globe of the Earth by J[ohn] Senex FRS'. But it so happens that John Senex died in 1740. And it so happens that after his death, his wife Mary Senex took over and managed, under his name, the prominent cartographic business he had developed. Because we know the date of the voyage marked on the globe, we know that it was published not by John, but by Mary. As proprietor, Mary Senex was active in scholarly and artisanal circles and even had a letter published in *Philosophical Transactions of the Royal Society*. Senex's decisions here, in ensuring her products were updated to reflect contemporary interests, have also enabled us to date, and thus identify, her work.

*A New and Correct Globe of the Earth*, John Senex, Mary Senex, about 1750

A New & Correct
GLOBE
of the Earth
By I.Senex F.R.S.

Tropic of Cancer

New
Mexico

THE PACIFIC OC
The Æquinoctial Line

The E     GREAT SOU

SCO

LIBRA

September

October

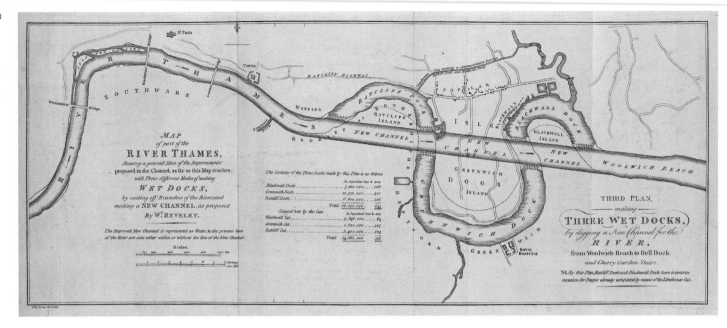

In the 1790s, architect and draftsman Willey Reveley (1760–99) proposed a radical new route into London, suggesting that the sinuous curves of the Thames be cut off by digging a straight channel through Rotherhithe, the Isle of Dogs and the Greenwich Peninsula. According to Reveley, such a plan had several advantages: the navigational challenges presented by the meandering Thames would be no more, and new wet docks would be created naturally in the cut-off curves. His plans were social engineering too. In a decade when, because of the war with France, recruitment to the Royal Navy was a pressing concern, he argued that creating islands would lead those who lived there to 'acquire *those Habits* which *lead* to the *Naval Service*'. Reveley's were not the only plans put forward for redeveloping the Port of London in the 1790s. They were, in fact, among a number of proposals sought as those with financial interest in international and colonial trade including the transatlantic trade in enslaved people, sugar and tobacco, worked to expand and transform the area. The aim was to increase the Port's capacity for shipping and better regulate labour on the river itself, where an informal economy flourished. When presented to Trinity House, Reveley's plans were judged to be 'novel, grand and captivating', but, in the end, disregarded as 'entirely impracticable'.[43]

*Map of part of the River Thames (C) Third plan making three wet docks by digging a new channel for the River...*, Willey Reveley, 1796

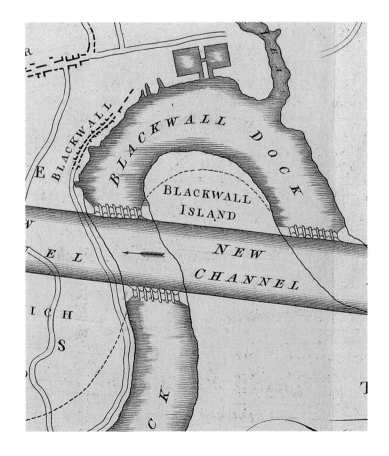

*The Road to Germany*,
unknown maker, about 2015

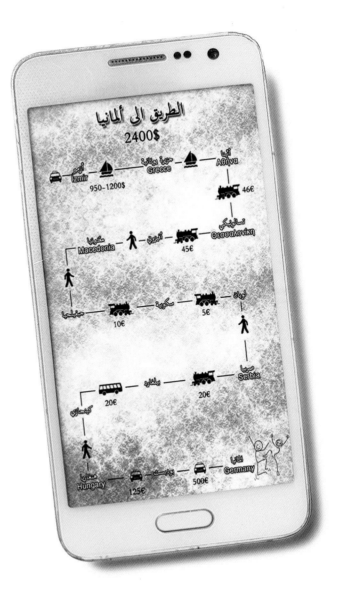

Shared tens of thousands of times among Arabic-speaking refugees
in 2015, this map, a small digital graphic, details the so-called
'West Balkan Route' from Izmir in Turkey to Germany, via Greece,
Macedonia, Serbia and Hungary. The map shows different stages of
the passage between capitals and border towns, the relevant modes
of transport and the currency required and likely cost. Arrival in
Germany is depicted by two stick figures, dancing in celebration. We
do not know who created this map, but they had detailed knowledge
of the route. Sent via WhatsApp and other social media applications,
its use would have been supported by other information, provided by
those who had already reached Western Europe, as well as agents,
handlers and smugglers. The map is an example of the sort of
ephemeral information people can be forced to rely on if the journey
they are making is deemed illegitimate by those with control over
borders. The information was precarious, too: when border fences
were erected in Hungary between late 2015 and early 2016, the route
described by this map became less relevant and fell out of use.[44]

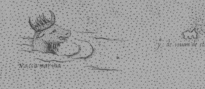

Vacca marina

y. della Afention

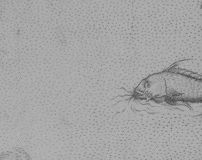

# S

## IS FOR

# SEA
# MONSTERS

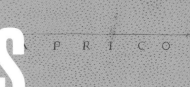

In the sixteenth century, it was a widely held belief that marvellous creatures inhabited the oceans, particularly the waters of the North Atlantic. Such thinking brought together different strands of knowledge. Classical scholars maintained that monsters roamed at the edges of the inhabitable world and that there were versions of everything on land in the sea. Biblical texts suggested the enormous beings in the deep – sea monsters – were created by God, even if they stretched and challenged human notions of nature. Meanwhile, parts of animals, unknown further south, were brought from northern waters to Europe by fishermen. Mysterious carcasses were washed up on beaches and word of them sent around European royal courts by letter. Where some maps offered description of the wonders and dangers of the seas, others included these creatures as decorative motifs to make their products attractive to buyers.

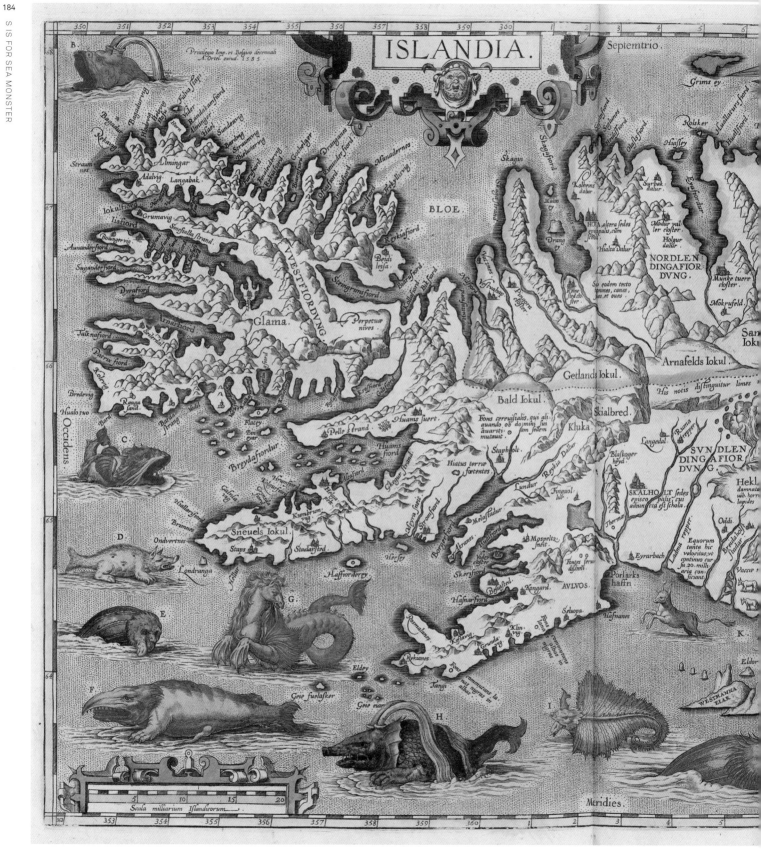

This map of Iceland not only depicts but also describes the marvellous creatures of the early-modern North Atlantic. On the back of the map is a key to each monster. To the top right is the 'navhal', with 'a tooth in the forepart of his head, standing out seuen cubittes', which was 'solde for Unicornes horn', supposedly an antidote to all poisons. In the bottom left is the seahorse with a flowing mane, which 'doth the fishermen great hurt and skare'. On the opposite side, in green, is the Steipereidur, 'a most gentle and tame kind of whale; which for the defence of fishermen fighteth against other Whales … no man may kill or hurt this sort of whale'. To the lower left is the 'sea-hogge', with dragon's feet and eyes in its belly. First reported in Rome in 1537, the creature was read as a portentous sign – indeed, warning – from God about the vanity of the age in which it appeared. The religious turmoil of the Reformation meant that there was a peculiar expectation of such signs from Roman Catholics and Protestants alike. Giving readers detail of the dangers, opportunities and purposes of those things that dwelt in the sea, the map was published at a time when there was great appetite for learning about and responding to the existence of monstrous creatures.

'Islandia', Theatrum Orbis Terrarum,
Abraham Ortelius, 1585

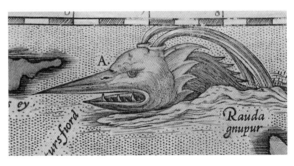

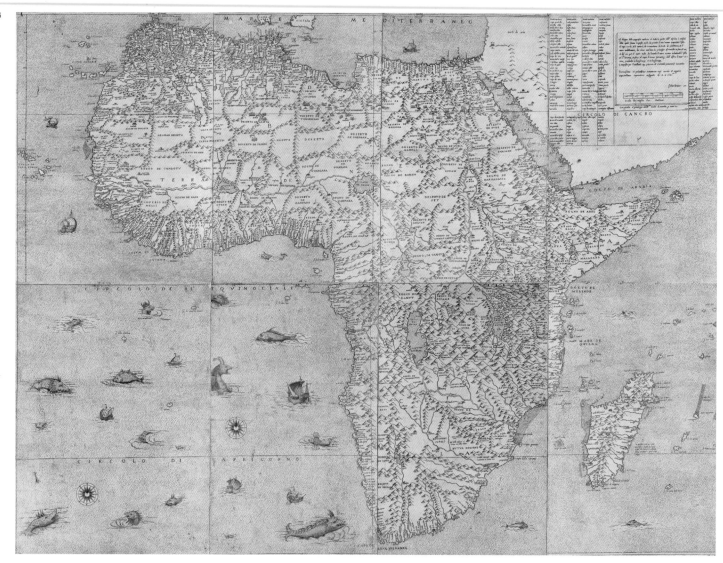

For many cartographers the inclusion of sea monsters in their work was more simply a form of decoration. At least half of this eight-sheet map of Africa by Venetian cartographer Giacomo Gastaldi (1500–66), including one whole sheet, is taken up by the sea. For potential customers, the addition of sea creatures made the work more attractive. In sixteenth-century Venice, Florence and Rome, printed maps were produced by the same people, sold by the same shops and often consumed in similar ways as figurative prints. What mattered to those purchasing was not necessarily the most up-to-date geographical information, but the most beautiful print. The creatures depicted drew on what were effectively stock images of sea monsters, including the 'Vacca Marina' or sea cow, shown here to the north-east of Ascension Island. If the alternative was engraving a whole sheet with the repetitive short lines used to represent the sea, it is perhaps unsurprising that these beasts populate Gastaldi's ocean.

*Il doegno della geografia moderna*
*de tutta la parte dell'Africa...,*
Giacomo Gastaldi, 1572

de iouan de steuam

vacca marina

p. della Asontion

ta. elena

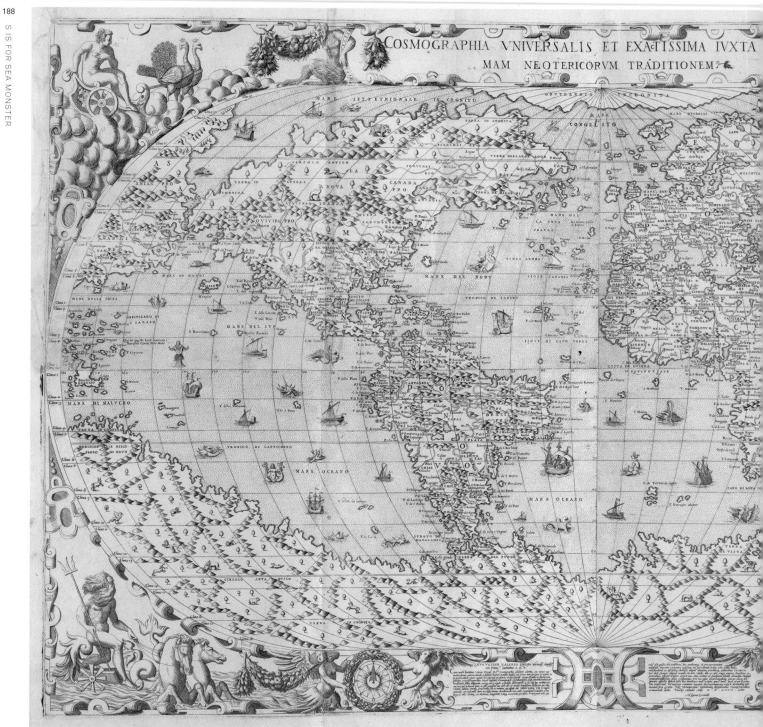

The sea-bishop and sea-monk – one with scaly mitre and the other with scaly habit – depicted on this map were reported sightings in the sixteenth century. The bishop, it was said, had turned up in Poland in 1531 and had an audience with the King, who was rather inclined to keep it. Instead, after successfully appealing to the other (land) bishops who were there, the sea-bishop was returned to the coast and, after making the sign of the cross, leapt into the water and swam away. This tale, though reported with some scepticism, was still included in sixteenth-century accounts of sea creatures. The sea-monk was found in the Oresund, the stretch of water between Norway and Denmark, and the Danish King, Christian III, sent news of it to all the courts of Europe. The sea-monk was ascribed greater legitimacy thanks to earlier textual references, including one by Albertus Magnus, a thirteenth-century scholar whose work on natural history was a key source for sixteenth-century writers. Sea-monk and sea-bishop turned up as decorative additions, too, including on this map. While the accounts of them related to the specific locations of the North Sea and the Baltic, Camocio placed them in the Pacific and Indian Oceans, where they serve more as curious decoration than images relating to particular sightings.[45]

*Cosmographia universalis et exactissima iuxta postremam neotericor traditionem*, Francesco Camocio, 1567

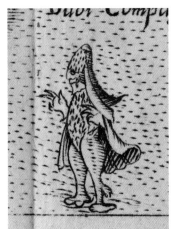

Not so much a map decorated with remarkable sea creatures as a remarkable sea creature decorated with a map, this object depicts a scene from the life of St Brendan, an Irish saint who lived about 484–577 CE. Accounts of his life, which were popular in medieval Europe, described how he set off from Ireland with a party of monks in search of the 'Promised Land of the Blessed', which became known as St Brendan's Isle, and travelled the Atlantic for seven years. Looking for somewhere to celebrate the Easter Mass, they came upon an island and some of the monks landed, lighting a fire for warmth. The island then moved and the monks fled. Brendan, who remained in the boat, informed the monks that the island was in fact a great fish named Jasconius, who had been startled by the heat of their fire. In subsequent years, Jasconius allowed Brendan and his companions to land on his back for Easter, as Philiponus shows here: the company kneels before an altar on the scaly spine of a great sea creature. The map shows the Iberian Peninsula, part of West Africa, the Canary Islands and the mythic St Brendan's Isle, which Brendan and the monks eventually arrived at, spending 40 days there before returning to Ireland.

'St Brendan in the Mid-Atlantic',
*Nova Typis Transacta Navigatio*,
Honorarius Philiponus, 1621

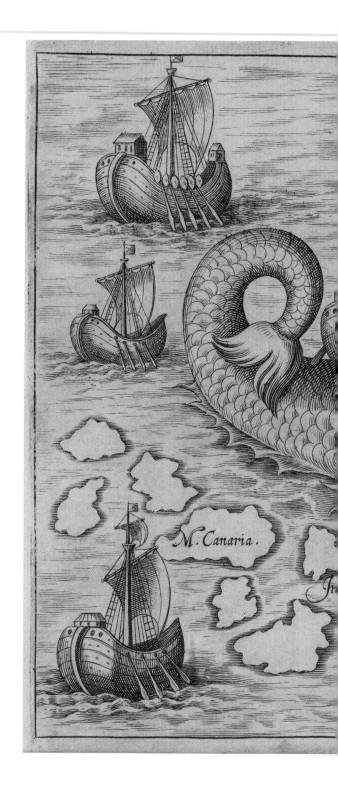

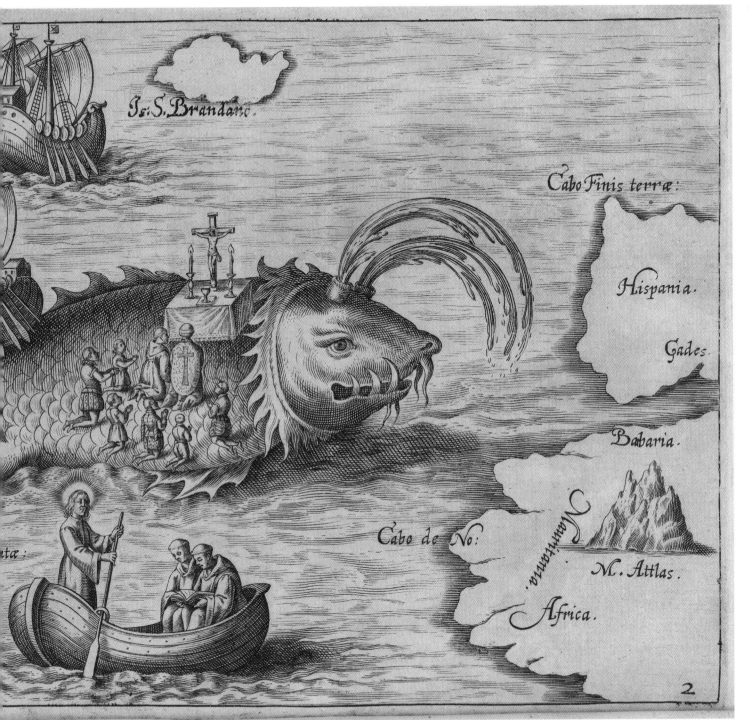

Is. S. Brandani.

Cabo Finis terræ:

Hispania.

Gades.

Babaria.

Mauritania

Cabo de No:

M. Attlas.

Africa.

2

**T** IS FOR

# TREASURE

Cemented in the popular imagination by nineteenth-century fiction, maps showing the location of buried pirate treasure are in fact inventions of the genre. Nonetheless, there are still many examples of treasure-related maps, among which are those that commemorate its capture or detail the routes by which it travelled. Maps can even be treasure in their own right, so guarded and valuable that they are themselves prized objects.

'Treasure Island', *Treasure Island*,
Robert Louis Stevenson, 1886

Robert Louis Stevenson (1850–94) was not the first author to tell a story about a treasure map, but the popularity of his novel *Treasure Island*, first published as a book in 1883, helped to fix the idea in the public imagination. Pirates had maps to hidden treasure, where 'X marks the spot'. According to Stevenson, the book had, in fact, started with the map, drawn during a game with his nephew. From there, the story of a map to pirate gold found in a dead man's sea chest developed; a voyage in search of buried riches infiltrated and hampered by pirates. Stevenson's original map was lost, so the one actually published and reproduced in copies of the book was reverse engineered from the tale. 'It is one thing to draw a map at random', Stevenson reflected, 'it is quite another to have to examine a whole book, make an inventory of all the allusions contained in it, and with a pair of compasses painfully design a map to suit the data.'[46] Within the book, although discovery of the map led to the pursuit of buried treasure, in the end X marked the spot where the treasure was not, having been moved sometime earlier. For both Stevenson and his characters, the map was inspiration and trouble. And, though without historical parallel, Stevenson's treasure map has had a huge influence on the way in which generations of people think about pirates and historic maps.

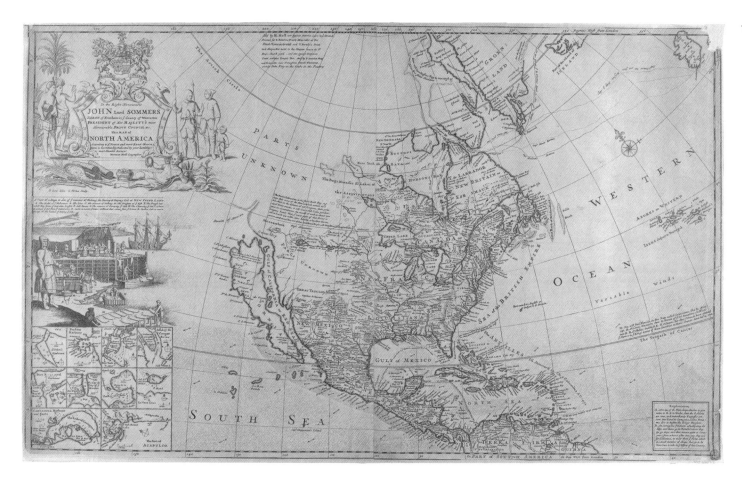

*Map of North America*,
Herman Moll, about 1712

From the middle of the sixteenth to the end of the eighteenth century, wealth extracted from the Spanish colonies in South America, often by the labour of enslaved people, was transported to Spain, or across the Pacific to the Philippines, by an annual treasure fleet. Silver, gold, indigo, cochineal, animal hides, precious stones and other valuable goods were all carried by vessels that sailed together under the protection of warships. This map of North America by Herman Moll shows, among other detail, the route taken by the fleet every year. In the Caribbean, vessels sailing from Santa Cruz in Mexico, and from Cartagena, in what is now Colombia, would meet at Havana, Cuba, before making the journey to Spain. It highlights the strategic importance of the Spanish control of Florida, as the whole fleet passed through the Gulf of Florida on its way into the Atlantic. Produced during the War of the Spanish Succession (1701–14), when, following the death of the childless Charles II, European states fought to determine the future dynastic control of Spain's imperial possessions, the map demonstrates contemporary fascination with South American wealth.

ring & Drying Cod at NEW FOUND LAND.
ishing. D. The Dressers of ye Fish. E. The Trough into
of Carrying ye Cod. H. The Cleansing ye Cod. I. A Press
od that comes from ye Livers. L. Another Cask to receive

C.

H.

The Baron Lahontan in his first Book
says that some of the Mozeemleck, na
cipal River empties it self into a Salt I
gues in circumference. the mouth of wich
Leagues broad. That ye lower part of th
adorn'd with six noble Cities. besides a hu
great an small. round that sort of Sea. and that ye Ri

Straits of Annian

Mozeemleck

High Mountains

Country

PARTS

UNKN

Many
Villages

Gnacs
res Co

Many Villages
on ye Islands

Morte

C. Blanco

C. St Sebastian
C. Mendocino

C. de Fortune

New Albion is laid down Accor
ding to the Observations maed by
St Francis Drake A.D. 1578.

NEW

C
A
L
I
F
O
R
N
I
A

High Mountains

GREAT TEC

Apaches de

GULF of CALIFORNIA or RED SEA

Po St Francis
Drake

Mount
Neuada

ALBION

M. St Martin

Canots

SEVO

Po de Monterey

R. de
Sardines

Gigate
Isle

Po Carinde

R. d Anguchi
Hopi

Maconabi
Logapapi
Anguamuchuari
Oldedba
Moatpic aiago
Panaria

R. of Good

Colorado

Mogu
Gua

Casagrande
Del. 1604.
S. Catalina

NEW

I. St Clement

I. de Parraros

S. Martin
Tuc
Addi

I. de Cuintas

la Conception
St Nicolas
St Isidore

Baipia

Son
Cristoval P.

C. St Apollina

Iuan
St Iuan

Gigantes
Edues

R. de la
Paz

Cenou P.

Wate la Gueda
Seralbo
Perka

Paz P.

Cotas

Ullao

S. Luca

Amblada

S. Thomas

J. Ma
C. Cor

la Nublada

La Roca
Partida

Fo Sp

SOUTH SE

Capt Chipperton's P.

Newark

York

Brook
land

Stats
Island

P. of Long
Island
Gravesend

Ashley &
Cooper River.

Settlements
upon Ashley
& Cooper
Rivers

Coney

Charles
Town

Old
Char
To:

New York

Wapoo

Sulli
vant I.

gada I.

Porta de la
Mare

Cast & I.
de S. Iuan
De lua

ye Road

he Bay of Po Bella

La Vera Cruz

Forta
Grande

Porto Bella

Mt Naqualla

Port Marques

South west Channel
Gri. ta I.

Naqualla R.

The Port of

In 1628, Admiral Piet Hein (1577–1629) of the Dutch West India Company successfully captured the Spanish treasure fleet off the coast of Cuba. For the young Dutch Republic, still at war with Spain in a long fight for independence, this victory was a great coup, celebrated in pamphlets, poems and popular songs. At the more opulent end of the commemorative scale there were medals, like this one commemorating Hein's success – portable, durable and easily reproduced – which offered a brief account of events and, typically, an allegorical image on the reverse. Here, the maker has chosen a map, centred on the source of Spain's wealth (incidentally, it is also one of the earliest known maps to show California as an island).[47] A verse from the Old Testament book of Jeremiah

surrounds the map and reveals the meaning behind the image. It translates: 'All the nations will serve him, his son and his grandson, until the time for his own country comes in its turn, when mighty nations and great kings will subjugate him.' Making clear the confidence and ambition of the Dutch Republic, the medal casts the capture of the fleet as a clear sign of the shifting balance between the new state and the Spanish Empire.

Medal commemorating Admiral
Hein and the capture of the Spanish
Silver Fleet, unknown maker, 1628

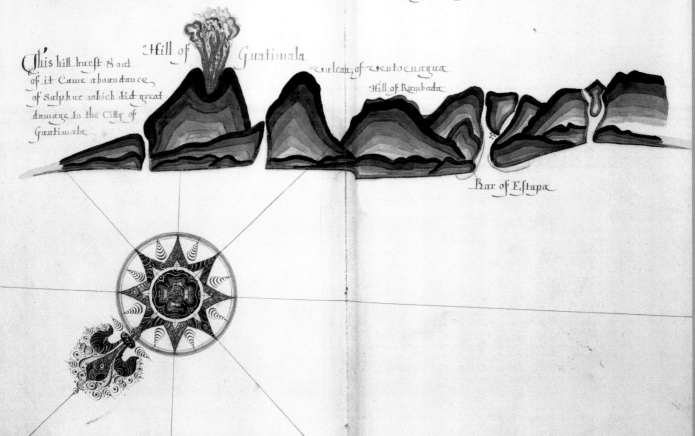

# GUATIMALA

This hill burst & out of it Came abundance of Sulphur which did great damage to the City of Guatimala

Hill of Guatimala

Vulcan of Vento curaqua

Hill of Rimbada

Bar of Estapa

From the hill of Guatimala to the bar of Estapa is 8 Leagues Coast runs NE & SW.

From the said bar to the River of Monticatca is 10 Leagues Coast runs NW & SE.

Sometimes, maps were treasure. In 1681, English buccaneer Bartholomew Sharpe (1650–1702) was sailing in the Pacific, seeking to capture some of the silver being taken from South America in Spanish ships. On taking the Spanish ship *Rosario*, Sharpe's crew threw some dull looking metal overboard, mistaking silver for tin. What they did not mistake, however, was the value of the *derrotero*, or book of maps and sailing directions, which the *Rosario*'s crew had not disposed of quite quickly enough (it was common practice to jettison navigational material). During a period in which the Spanish kept very close control over the navigational information that gave them access to the ports of South America, this volume was a true prize. On return, Sharpe was to be tried for piracy because of his exploits, but the trial was cancelled. It has been suggested that the importance of the captured book was such that Sharpe's other activities were overlooked. Taken back to England, the work was copied multiple times by renowned London cartographer William Hack (active 1671–1702). One of 149 maps in this volume, this page shows part of the coast of Guatemala, with a description of a volcanic eruption as well as bearings and distances between different places.

'Guatimala', *A Waggoner of the South Sea*, William Hack, 1685

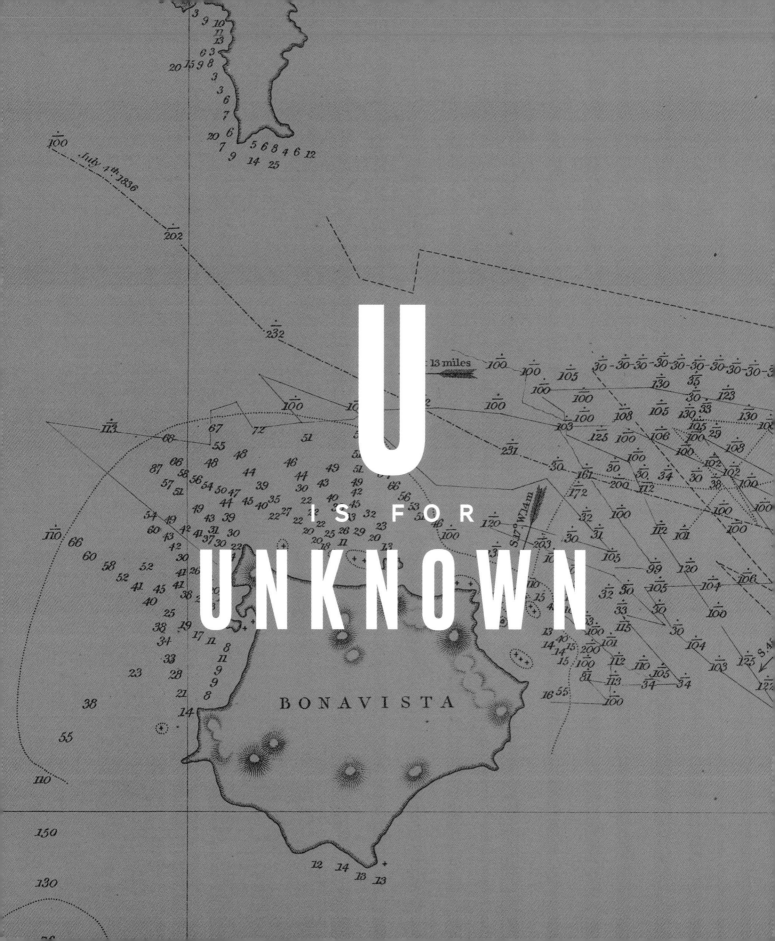

U

IS FOR

UNKNOWN

BONAVISTA

The topic of the 'unknown' necessarily raises the question 'unknown to whom?' For the tens of thousands of people who had lived for hundreds of thousands of years on the island we now know as Australia, for example, the land was clearly not unknown, even if it was to European map-makers and consumers until the later eighteenth century. Casting places as unknown can have far-reaching consequences and reinforces the idea of a land being available for conquest, with devastating effects on the people that live there. As readers, we must be careful too – in considering the unknown on maps we risk viewing ourselves as superior, now that we, after all, possess knowledge of the whole geography of the far side of the Moon, or the shape of the Antarctic continent. What is interesting is the way the unknown is dealt with. Our responses can take the form of speculation, blank space, intensive searching and reflection on how to interpret and work with uncertainty.

World map, Joan Martines, 1572

Cartographers in sixteenth-century Europe drew on a variety of sources when constructing their maps. More recent reports from mariners, merchants and other travellers supplemented and developed the geographical knowledge received from classical and medieval texts. Such learning came from philosophical reflection as well as from observation and this was especially true of the lands that cartographers described towards the Poles. In common with many European world maps from the sixteenth to the eighteenth centuries, this map by Joan Martines (active 1556–87), a cartographer who worked in Messina, Sicily, features an extensive landmass around the South Pole, labelled 'Terra Incognita' ('Unknown Land'). Extending

maps is in part thanks to a hypothesis made by the Greek philosophe Aristotle. He suggested that the Earth needed to be balanced and so any land mass in the northern hemisphere must have an equivalent in the south, 'for otherwise', as one sixteenth-century writer explained 'the stability of the world in its central position could not last'.[48] And so it was that this reasoned continent was combined with various stories of southern lands: Ophir and Tarshish, mentioned for their wealth in the Hebrew scriptures but not identified with known places Locach, or Beach, reported by Marco Polo after his travels to the East. The idea of the southern continent as a great unknown land, formulated according to textual scholarship and study, would go on to

In October 1959, the robotic Soviet spacecraft *Lunik 3* made the first ever photographs of the far side of the Moon. It successfully transmitted to Earth 17 of its 29 images, taken over the course of 14 orbits of our nearest celestial neighbour, in one of the biggest space-science coups up to that point. Covering only about 70 per cent of the Moon's far side, the images were grainy and difficult to interpret. They suggested a heavily cratered surface without the vast lava plains or 'seas,' familiar as the dark blotches on the Moon that are visible with the naked eye. Using data from the *Lunik 3* photographs, this globe was the first to represent the far side of the Moon and was made to advertise the scientific prowess of the Soviet Union. Published in 1961, it coincided with an intensification of the Cold War Space Race between the USA and the USSR, which saw then President John F. Kennedy announcing his intention to put a person on the Moon within the decade, and the first human space flight by Soviet cosmonaut, Yuri Gagarin. The globe itself featured a blank segment representing a lack of data and, as the Space Race gathered pace, it went out of date shortly after it was released. Regardless, it was both a celebration and a statement of future commitment, showing the dramatic success of *Lunik 3* and the work on lunar cartography still to be done.

Lunar table globe, Central Science
Institute, Moscow, 1961

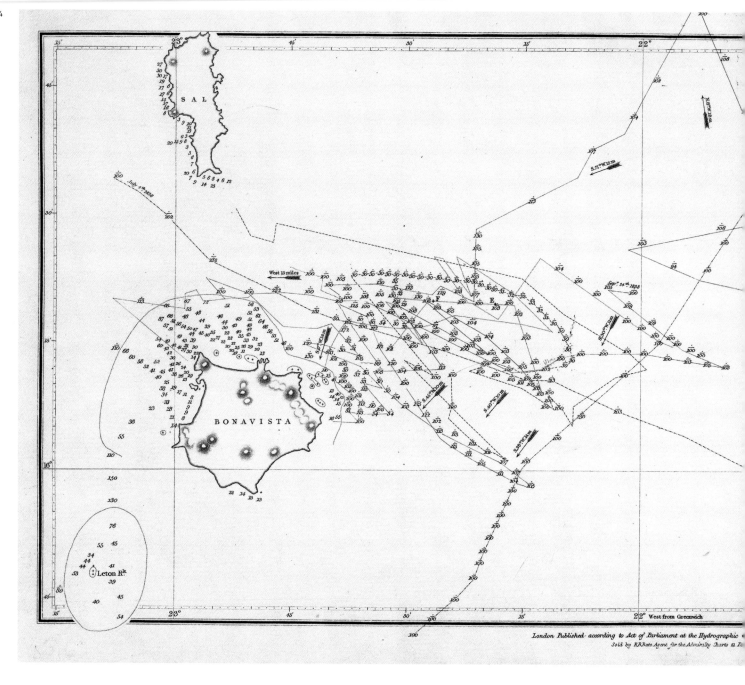

London Published according to Act of Parliament at the Hydrographic
Sold by R.B.Bate Agent for the Admiralty Charts 21 Po.

In September 1838, Captain Alexander Emeric Vidal (1792–1863) of the Royal Navy went in search of a rock near the Cape Verde islands in the Atlantic Ocean. It was held responsible for several maritime accidents, most recently the loss of the merchant vessel *Madeline* in 1835, and could be found in charts and sailing directions relating to the west coast of Africa. The only problem was that no one quite knew where it was. One book of sailing directions suggested it was 'somewhere to the Eastward or E. N. E. of Boa Vista'. One navigator placed it 42 leagues (or 145 nautical miles) from Boa Vista; another, 31 (107 nautical miles). As a writer for *Nautical Magazine*, which commented on naval and particularly navigational affairs, described, 'what with modern stories and ancient ones, and reckonings of all description ... the matter is very nicely cooked up for the edification of seamen and the glorious transactions of the insurance market, by all parties'.[49] Vidal had departed with six possible locations for the rock, marked A to F on this chart. After criss-crossing back and forth, diligently plotting the route of the vessel, he concluded that the Bonetta Rock simply

Tracks of H.M. vessels 'Etna'
and 'Raven' September 1838
in search of the Bonetta Rock,
Hydrographic Office, 1841

205

U IS FOR UNKNOWN

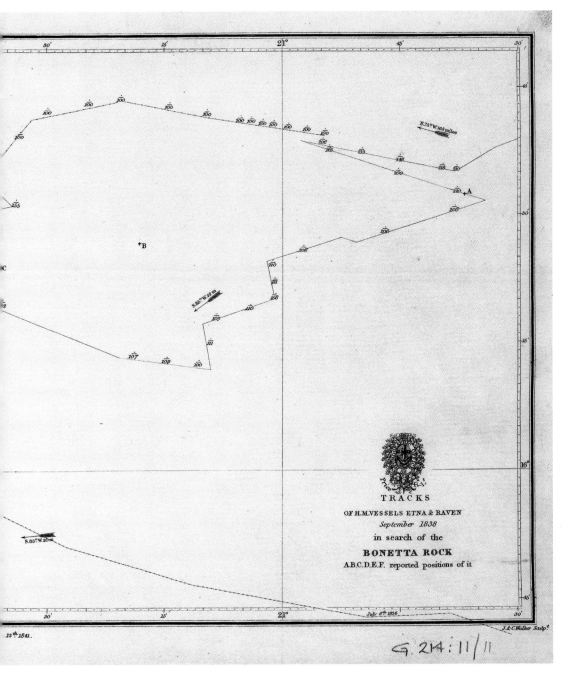

did not exist. It was instead a mis-plotting of reefs extending from
the island of Boa Vista, which had resulted from errors, caused
by strong currents in the region, in dead reckoning, those rule-of-
thumb methods for calculating distance and direction so crucial to
navigation. Published first in the 1839 edition of *Nautical Magazine*
and then in 1841 as a separate chart issued by the Hydrographic
Office, Vidal's work served to demonstrate both the non-existence
of the sunken rock and the navigational demands of sailing near
the Cape Verde islands.

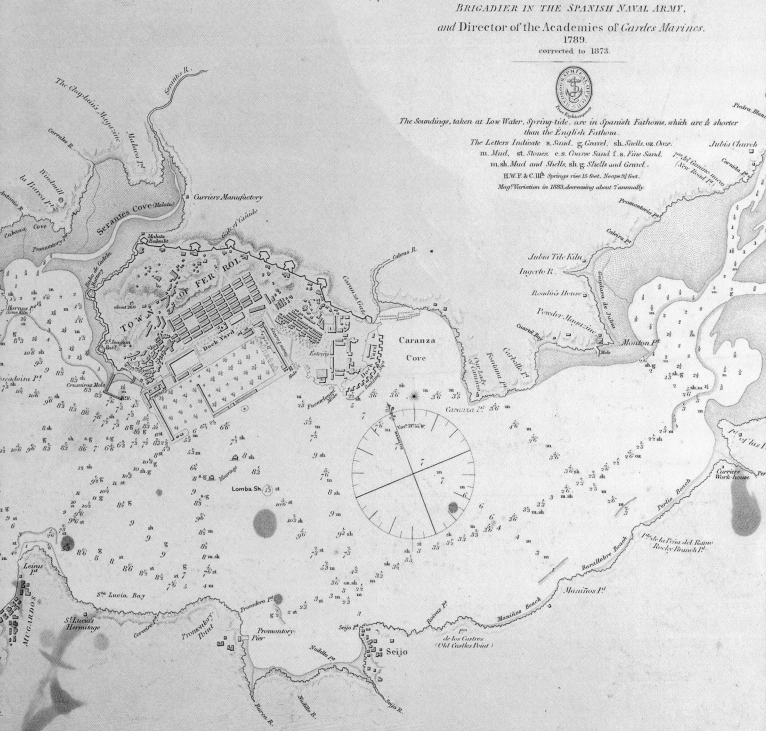

# PLAN
## OF THE MOUTH AND HARBOUR OF FERROL

Palma Castle Lighthouse Lat. 43° 27′ 45″ N. Longitude 8° 16′ 8″ W. of Greenwich

*Geometrically Surveyed*

### By DON VINCENT TOFIÑO DE S. MIGUEL
#### BRIGADIER IN THE SPANISH NAVAL ARMY,
*and* Director of the Academies of *Gardes Marines.*
1789.
corrected to 1873.

The Soundings, taken at Low Water, Spring-tide, are in Spanish Fathoms, which are ½ shorter than the English Fathom.

The Letters Indicate   s. Sand.   g. Gravel.   sh. Shells.   oz. Ooze.
m. Mud.   st. Stones.   c.s. Coarse Sand.   f.s. Fine Sand.
m. sh. Mud and Shells,   sh. g. Shells and Gravel.

H. W. F. & C. III.ᵘ Springs rise 15 feet, Neaps 9½ feet.

Mag.º Variation in 1883, decreasing about 7′ annually.

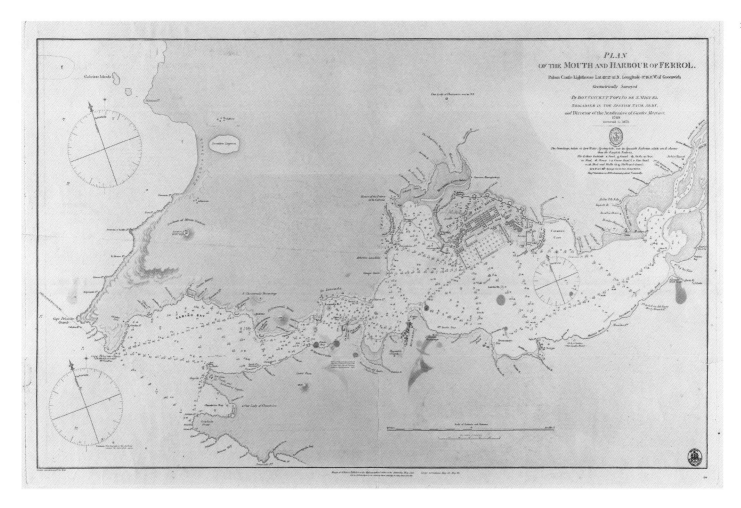

At the end of the nineteenth century, an argument spilled out into the pages of *The Times* about how good navigators had to be sensitive to the unknown in navigational charts. In 1892, the battleship HMS *Howe* had been badly damaged after striking a reef at the entrance to the harbour of Ferrol, on the north-west coast of Spain, which was unmarked on this chart produced by the Admiralty Hydrographic Office. For one correspondent, this was a sign of intolerable carelessness on the part of the Hydrographic Office. For those involved in the work of hydrography, however, it was evidence that navigators were misinterpreting blank spaces in between areas of soundings (depth measurements) on navigational charts. Far from being proof of deep water, such areas in fact indicated where survey work had not been done and therefore needed to be read as unknown and potentially dangerous, as opposed to known, deep and safe. The previous year, the Admiralty had published a pamphlet, *Notes Bearing on the Navigation of HM Ships*, which dwelt on how charts as navigational tools should not be granted too much authority by their users and how navigators had to be sensitive to when and the way in which they were produced in order to understand the reliability of the data. The argument caused by the *Howe* incident was, in the end, about expertise in navigation and recognising what remained unknown, even amid the intensive regime of survey work developed in decades past.

*Plan of the mouth and harbour of Ferrol*, Hydrographic Office, 1886

V

IS FOR

VIEW

nitola.

Tre Fontani Tower.

An Algerine

APPEARANCE OF TI

No one view or perspective holds the key to navigation. For seafarers past and present a variety of different views help to develop a sense of the coastlines and oceans across which they travel. That means that a human's eye view and a bird's eye view – a profile and a map – are both helpful images. A coastal view, the appearance of a coastline from a ship, was a vital element of coastal navigation even before the introduction of charts. Marine views are also an artistic genre in their own right. As in many areas of navigation, there is much overlap between the artistic and the technical. Seafarers frequently sketched without practical intent; on occasion artists were employed to make navigational views.

And also in the Golf of Venice to Corfu.

West and by south 8 leagues.

The cape Cyta Nova.

Ancona.

In this manner appeareth the corner & the city of Ancona, being over against it, & the Mountaine of Cita Nova south from you.

Ancona appeareth thus, when the Mountaine lies w. n. w. from you.

As you saile by the hig land of Ancona, it appeareth in this manner.

Ancona west and by north 6 leagues.

Ancona s. w. and by west 8 leagues.

Ancona s. s. w. 9 leagues.

Ancona s. w. and by south 8 leagues.

In this manner appeareth the City Chiosa, being about 4 leagues from it, and it lies five leagues south of Venice.

Thus appeareth the towers of Venecia, when you are about 3 or 4 leagues from land, you cannot then see it by reason of the low land.

Chiosa.

When you come neare to Chiosa, it appeareth in this manner, there is northward of the city a river, but not that by which you sail op to Venice.

These houses stand to southward of the river of Venecia, and to northward of that is the river to Malamocco.

Sea atlases frequently combined charts, textual sailing directions and coastal profiles. In a work like Arnold Colom's atlas of the Mediterranean, charts printed from copper plates were accompanied by pages of text interspersed with woodcuts showing the shape of a coast as seen from the sea. Indeed, Colom's estimation of their importance meant that this volume of 124 pages and 19 plates has 54 pages of coastal views. The land is shaded with diagonal lines and the images also show notable buildings and trees accompanied by explanations. This page of views takes the reader from Ancona on the east coast of the Italian peninsula, past Venice to Istria on the coast of what is now Croatia. Rather than presenting the whole coastline, it shows what were thought to be the most significant places. Describing the view of and maritime approach to Venice, the text explains, 'these towers stand in the city of Venice, when you first see them, you are yet four leagues from land, which is very low, in so much that you see the towers before you see the land'. While Venice's economic power was in decline by the seventeenth century, it remained a major seaport and details about the approach to the coast were vital for navigators.

'Golf of Venice to Corfu', *Lighting Colom of the Midland-Sea*, Arnold Colom, 1660

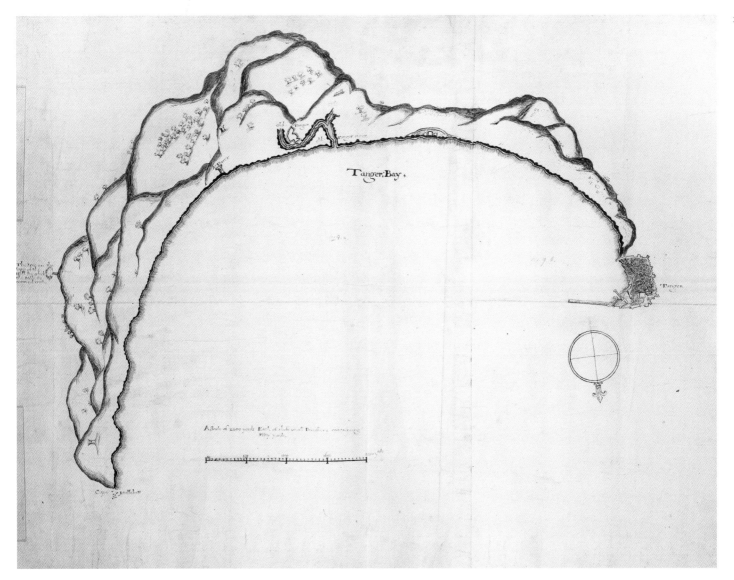

Sometimes cartographers combined a coastal view with the chart itself, resulting in a curious sort of panorama or rotational view. The maker of this chart of the Bay of Tangier on the North African coast, did just that, drawing the hills as seen from the sea arrayed around the crescent-shape of the bay viewed from above. The map was one of many English representations of Tangier produced in the later seventeenth century and was made during the short period in which the city was part of the English colonial empire. Tangier had been part of the dowry received by Charles II on his marriage to the Portuguese princess Catherine of Braganza in 1661. Charles decided it should be a naval station and an English commercial centre – a challenge to the Dutch advantage in the region and a link between the Atlantic and Mediterranean worlds. In just over 20 years, around £2 million was spent by the King and his ministers on developing the harbour at Tangier, more than was invested in any other colony.[50]

By 1684 the English had left, demolishing administrative buildings and costly harbour infrastructure as they abandoned them. The port was not big enough for large naval vessels, so never became a home for an English fleet. It had remained commercially unimportant, despite high hopes for its profitability. The large number of Catholics posted to the region ensured that it was viewed with suspicion in a Protestant empire. Furthermore, it suffered an increasing number of concerted attacks by Moroccan forces. It was in this context that the English Parliament refused to pour any further funds into the city. This map belonged to Royal Navy Admiral George Legge, 1st Baron Dartmouth (c.1647–91), sent by Charles II to evacuate the failed colony in 1684. The houses, fortifications, city walls and harbour depicted on it were all blown up before the English left Tangier.

*Tanger Bay*, unknown artist, 1662–80

The navigational value of coastal views was such that in 1799 the British Admiralty engaged a marine artist, John Serres (1759–1825), to sail with the Channel Fleet. His principal task was to produce sketches of the French harbour of Brest, desirable for strategic purposes, but he went on to make views of the coastline of the Channel between southern England and northern France, as well as the coastal regions of southern France and Spain. Apparently encouraged by the captain of one of the ships he sailed on, he then began work on an English edition of a French volume of charts and sailing directions by René Bougard first published in 1684. Serres described how the captain had told him 'that if I were to give a version of it, I should be doing an essential service to my country, and that if the appearances of the Headlands and Charts were equally well executed as the descriptive parts were just and accurate, it would be one of the most valuable works ever published'.[51] Serres re-drew all the coastal views for the volume, which was published as *The Little Sea Torch* in 1801. In this series of views of Gibraltar, made from different angles, he used the arrangement of seabirds in the sky to indicate which feature was which, as described in the key. Although the charts, views and sailing directions were quickly superseded, Serres' output remained valued for its artistic merit. Even on publication, the periodical *The British Critic* proclaimed that its elegance 'renders it valuable even to those who have no occasion for the work as a guide to avoid [the shore's] dangers'.[52]

'Views of Gibraltar', *The Little Sea Torch*, J.T. Serres, 1801

Engrav'd for I.T. Serres's Little Sea Torch, and Pub.d by him, London 1801.

THE MOLE OF GERGENTI, AS SEEN FROM THE TEMPLE OF THE VIRGINS.

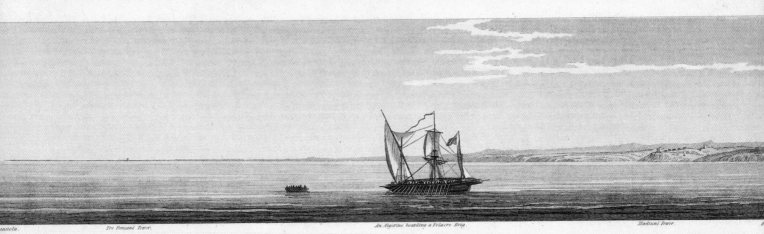

*anitola.*      *Tre Fontane Tower.*      *An Algerine boarding a Polacre Brig.*      *Madiuni Tower.*

APPEARANCE OF THE S.W. POINT OF SICILY.

PANTELLARIA, W.ᵗ S. 4 OR 5 LEAGUES.

This view, from *The Hydrography of Sicily, Malta, and the adjacent Islands* by marine surveyor William Henry Smyth (1788–1865), is of negligible navigational use. It shows a view from the land and not the sea. The Mole, or breakwater, of Girgenti (Agrigento), named in the title, is only just visible as a slightly heavier line on the water, jutting out from the land on the far left of the image, just below the horizon. The view is actually of the Sicilian temple, more commonly known as the Temple of Juno, with surveyors, including perhaps Smyth himself, at work beside it in the foreground. While surveyors produced views for practical purposes, they also drew for pleasure, to improve their eye and in ways that were influenced by and played with wider artistic traditions. This example self-consciously takes up the iconography of images from the Grand Tour, in which young,

aristocratic tourists were depicted conversing, relaxing or otherwise posing among classical ruins. Here such imagery is echoed by the naval officers at work. Often from more middling backgrounds, and at sea from a young age, naval officers had different routes into classical knowledge. Smyth himself had gone to sea aged 14 and had become a keen antiquarian as well as a surveyor. This view makes a claim for his expertise in both areas. While of little use for sailing into Girgenti, the view represents a different sort of navigation: through a world where classical knowledge was valuable cultural currency.

'The Mole of Gergenti, as seen from the Temple of the Virgins', *The Hydrography of Sicily, Malta, and the adjacent Islands*, William Henry Smyth, 1823

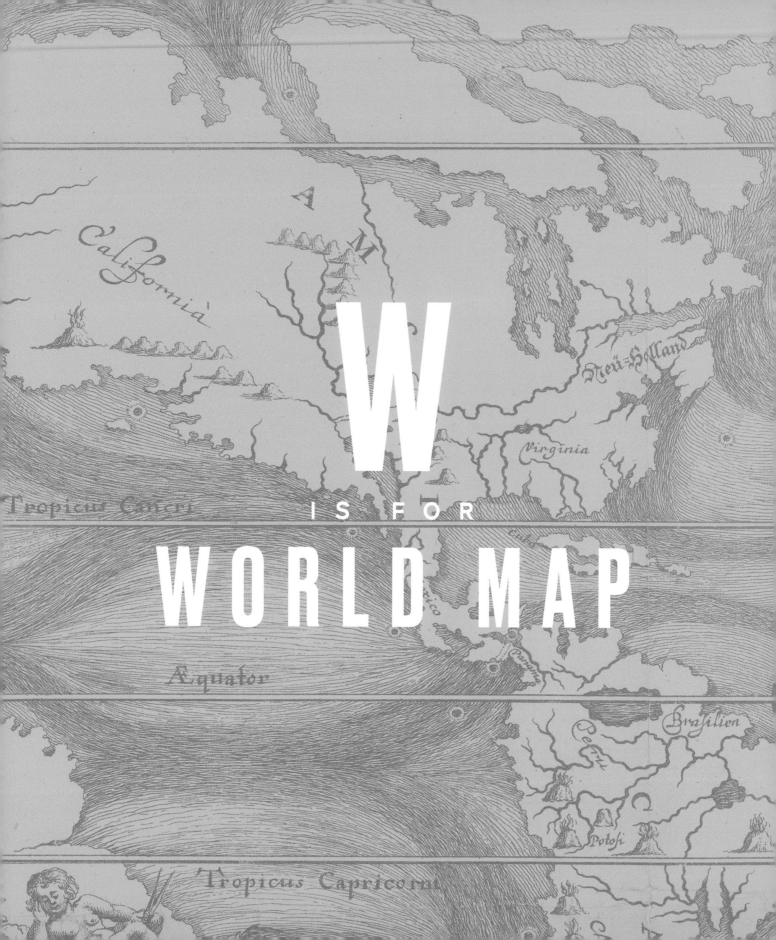

**W** IS FOR
**WORLD MAP**

European world mapping changed radically in the fifteenth and sixteenth centuries. The adoption of techniques to plot global geography mathematically along with proliferating ideas about how to represent the spherical Earth in two dimensions marked a new era in cartographic practice. Recognising that the curved surface of the Earth would always distort either the shape or the relative size of landmasses, or both, geographers played with different approaches to map projection. In the centuries following, and particularly from the nineteenth century on, world maps also became a useful way to visualise global phenomena, particularly to communicate with an increasingly map-literate public.

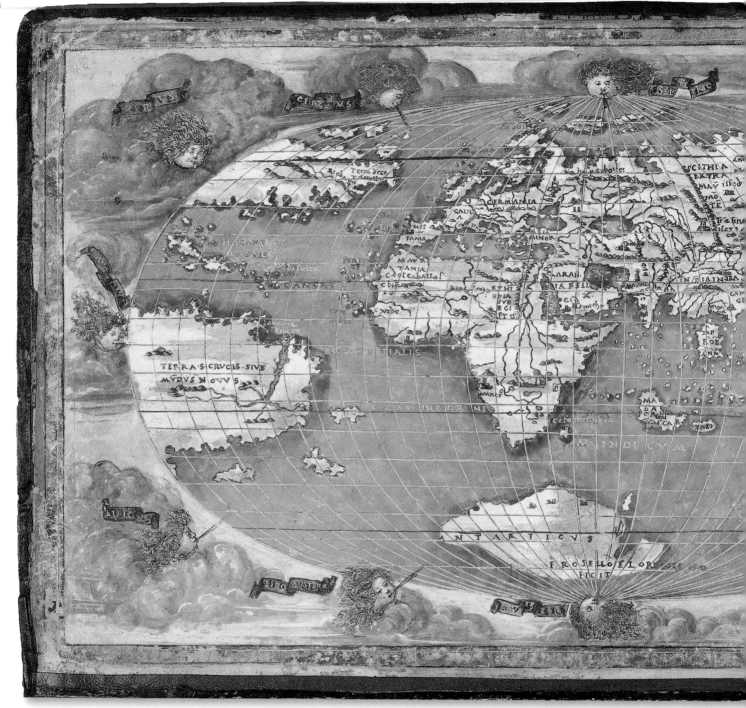

World map, Francesco Rosselli, 1508

Small and exquisitely coloured, this map was made by Francesco Rosselli (*c*.1448–before 1513), a Florentine miniaturist, cartographer, engraver and print-seller. It was produced during a period in which long distance voyages were substantially changing European geographical knowledge. In 1502, Christopher Columbus sailed on his fourth voyage across the Atlantic, seeking a route to East Asia. He believed he had made it when he arrived on what was, in reality, the Central American coast. It is this (mis)understanding that Rosselli depicted; place names from the east coast of Central America are marked on the east coast of Asia. To the north, Newfoundland, Canada, is part of the very east of the Eurasian continent and what we now know as South America is a large island labelled 'Terra S. Crucis, Sive Mundus Novus' (Land of the Holy Cross, or the New World). As well as recording a very specific moment of European geographical history, this is the earliest known map to include 360° of longitude and 180° of latitude within an oval projection. Rosselli, who had a keen interest in mathematics and was described as a cosmographer by contemporaries, was participating in ongoing experimentation about how best to represent the globe on paper. The oval form would later be adopted by other prominent sixteenth-century cartographers, including Giacomo Gastaldi and Abraham Ortelius.

This map by Ottavio Pisani (born 1575), an Antwerp-based, Neapolitan mathematician and cartographer, proposed a unique way of constructing a map of the world. Centred on the South Pole, the North Pole is the circular map's whole circumference as though the world has been splayed outwards from the south. Imagining a map laid out across the inside of an umbrella, the landmasses are all inverted east to west: Madagascar is to the left of Africa; Brazil is on the left side of South America. The latitudes of the southern hemisphere are carefully plotted, with degrees of latitude increasing in size towards the Equator. North of the Equator,

towards the edge of the map, degrees of latitude are incrementally compressed, meaning that closer to the North Pole the landmasses are very squashed north to south, as well as being very stretched east to west. Pisani was aiming for innovation, though, rather than practicality. As he wrote to Galileo Galilei, one of his correspondents, he had put 'within a circle the whole of the flattened globe, something that no one has done before'.[53]

*Globus Terrestri Planispherius*,
Ottavio Pisani, 1610

Grün=Landt

Island

Nor=wegen

Schweden

Califormia

A
M
E
R
I
C
A

Neu=Holland

Virginia

Tropicus Cancri

Meridian

Spanien

Rom

Barbarei

Libyen

A F

Æquator

Mexico

Peru

Brasilien

Potosi

Tropicus Capricornus

Chili

Süd=Land

Die W und Fluth
auff einer flachen
Landt=Karten
fürgestelt

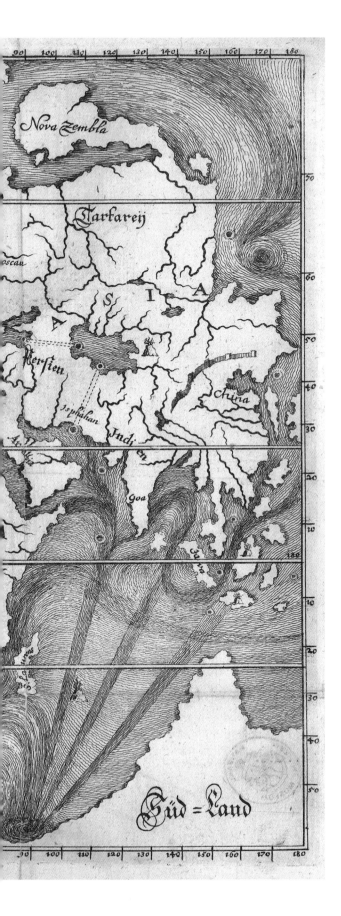

This map of the world illustrates a theory developed by German polymath Athanasius Kircher (1602–80), which linked the global movement of water with the existence of subterranean fire. Arguing that there had to be some sort of outlet for all the water that flowed into the sea, as well as a way of replenishing the rivers themselves, he proposed an underground system. There were, he suggested, chasms in the seabed into which water flowed, forming whirlpools. Once underground, water – driven by the heat of the earth and capillary action (the upwards movement of water through porous surfaces), and by the surface phenomena of wind and tide – flowed through channels, was pushed up to the mountains where it emerged as springs, and would then flow once more into the sea. This particular map was published in the illustrated periodical, *Relationes Curiosae*, a collection of curiosities published by the prolific German writer Eberhard Happel (1647–90). In the intellectually flourishing city of Hamburg, Happel's publication was the first to include contemporary scholarship in entertaining form and in the German language (rather than in Latin). The map features swirling currents, flaming volcanoes, whirlpools marked by dots in the ocean and even the underground channels that Kircher thought conveyed water between, for instance, the Mediterranean and the Red Sea. Kircher's theory was now available in visual form to a wider, though still scholarly, audience.

*Die Ebbe und Fluth auss einer flaschen Landt Karten furgestelt*, Eberhard Happel, 1685

Marie Tharp and Bruce Heezen's mapping of the ocean floor changed the way scientists thought about the bottom of the sea. Working at the Lamont Geographical Observatory at Columbia University in the 1950s and 1960s, Tharp (1920–2006) – not permitted to go to sea by her employer because she was a woman – plotted the sounding data that Heezen (1924–77) collected using new, continuous sounding technology. What she found as she compiled depth profiles made on vessels crossing the Atlantic was astonishing: a cleft running down the Atlantic, which became the evidence that confirmed the theory of continental drift, until then widely regarded as folly. Tharp employed a technique known as 'physiographic mapping' (representing physical features obliquely rather than with contour lines) in her map of the Atlantic. She and Heezen used the same system for their world map, which combined the results of their research and made it available to a wider audience. The choice of physiographic mapping was initially taken because of Cold War-era secrecy around sounding data. Any publication with bathymetric information was, in the decades following the Second World War, automatically classified in the USA because of its importance to top secret submarine navigation.[54] As Tharp reflected much later, though, 'our choice of map style turned out to be significant because it allowed a much wider audience to visualize the seafloor'.[55] Demonstrating something that had not been seen before, Tharp and Heezen's work radically changed both scientific and, thanks to maps like this, popular understanding of the nature of the ocean floor.

*The Floor of the Oceans*, based on bathymetric studies by Bruce C. Heezen and Marie Tharp, painted by Tanguy de Rémur, 1980

Mercator Projection 1 : 48,000,000 at the Equator.
Depth and Elevations in Meters.
Painted by Tanguy de Rémur.

copyright © 1980 Marie Tharp

# THE FLOOR OF THE OCEANS

*Based on Bathymetric studies by*
Bruce C. Heezen and Marie Tharp
*of the Lamont Doherty Geological Observatory*
*Columbia University Palisades, New York, 10964*
SUPPORTED BY THE UNITED STATES NAVY
OFFICE OF NAVAL RESEARCH
Third Edition · Third Printing

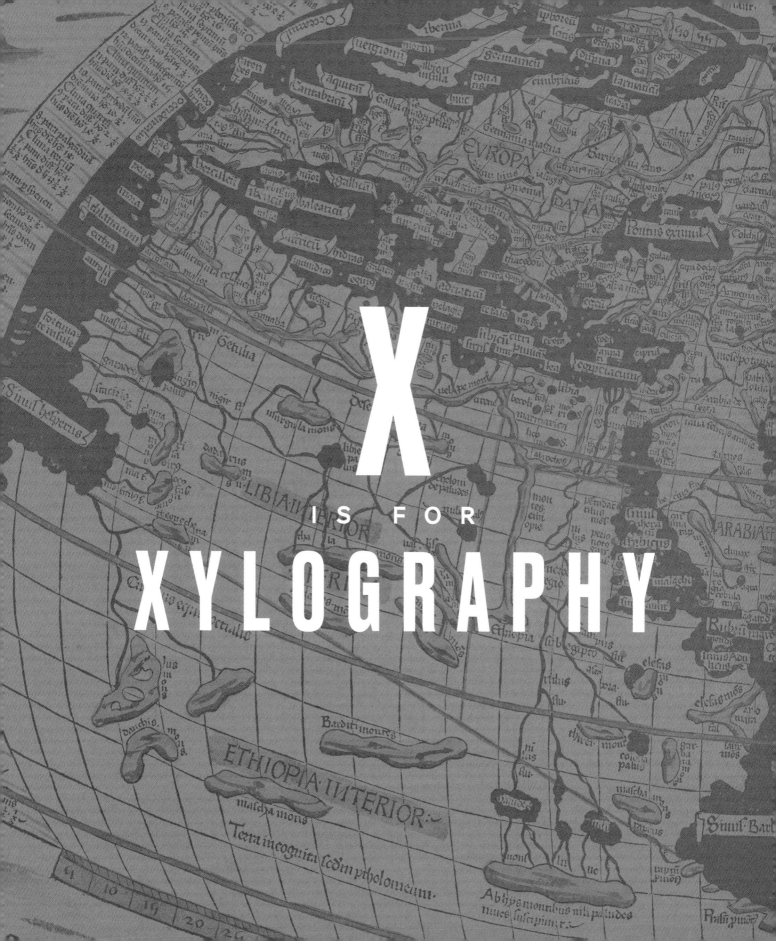

X

IS FOR

XYLOGRAPHY

Xylography, literally 'drawing from wood', refers to the art of cutting and printing from woodblocks. In Europe, the earliest printed images were made using this method in the fourteenth century, while in China the technique had already been in use for around 1,000 years. Woodcut techniques were particularly popular with European map-makers in the fifteenth and sixteenth centuries. Making an image in this manner involves transferring the map (in reverse) onto the cross-grain of a block of wood, then removing the parts of a block that should be blank in the finished picture with a pointed knife or, when removing large sections of wood, a chisel. The raised surface is then inked in order to leave an impression on paper, a process known as relief printing.

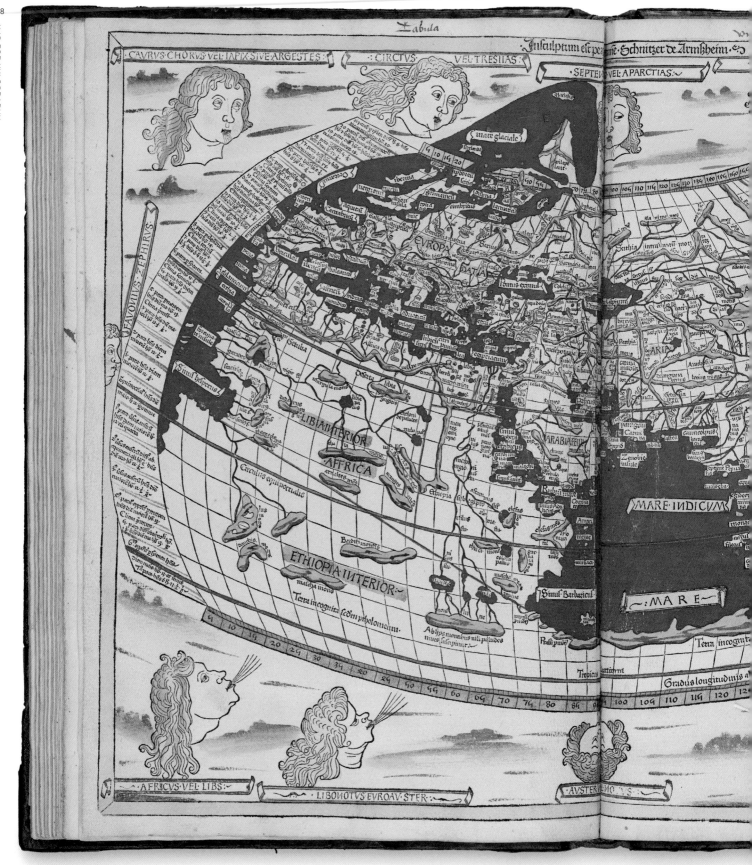

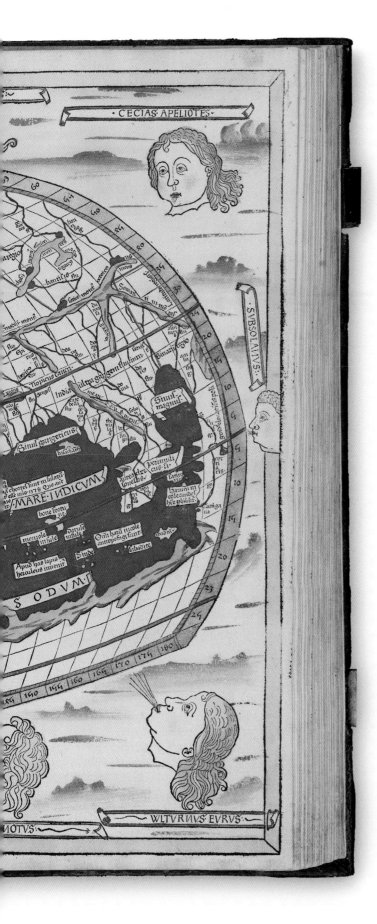

In the early 1480s, Leinhardt Holle (active 1480s), newly in possession of a printing press in the city of Ulm, Germany, hoped to capitalise on the strong Renaissance interest in classical learning and mathematical geography. He decided to produce a copy of the *Geographia*, a text by the second-century Greek-Egyptian scholar Claudius Ptolemy, which outlined both methods of map projection and the presentation of geographical data according to a mathematical grid. Holle spared no expense. In particular, the maps were woodcuts of a size never before seen in a printed book. They were such an impressive labour that, very unusually, the name of the person who cut the woodblock, a Johannes Schnitzer ('the Carver') of Armsheim, was included at the top of this map. The map also features a backwards capital 'N'. Because of this, and the fact that the image is signed by Schnitzer, historians have tentatively identified other unsigned works that include this symbol as Schnitzer's own. Holle's ambition, however, went beyond his means and his market. Printing with moveable type was still a very new enterprise and those in search of scholarly works had many reasons to prefer manuscript. Many printers quickly went bankrupt and Holle was no exception. The first Ptolemaic atlas to be published with woodcut maps, printed on fine paper imported from Milan, and often coloured using expensive pigments, ultimately put Holle out of business.

World map, *Claudii Ptolemaei Cosmographia*, Nicolaus Germanus and Johannes Schnitzer, published by Leinhardt Holle, 1482

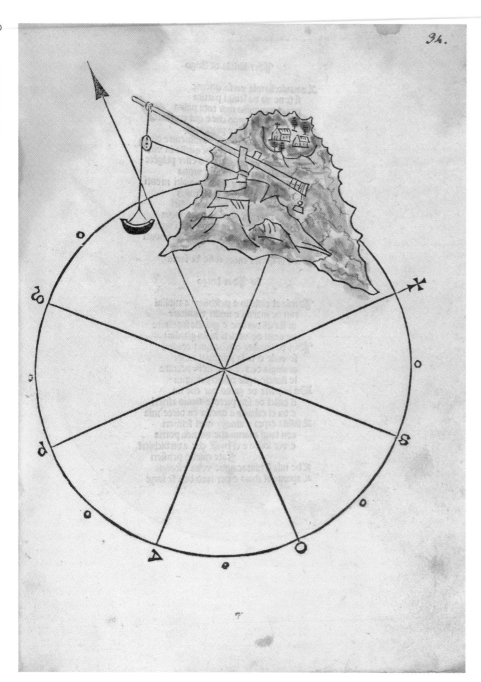

‘Caloeiro’, *Isolario*,
Bartolommeo Sonetti, 1485

The simple woodcut maps of islands in the Mediterranean in this *Isolario*, or Island Book, are accompanied by verse (barely visible through the page) in Venetian dialect. They are copied from manuscript work by fifteenth-century writer Bartolommeo Zamberti, known as ‘Sonetti’ because of his poetic style. A sea captain from Venice, he described in the book's preface how he had ‘stepped repeatedly on each isle … and with a stylus marked true position on the chart’. One of the features Sonetti introduced in his volume was the centring of a compass rose on the islands he depicted. The compass points take their names from the winds: *Tramontane*, the north wind, whose name derives from the Latin word for ‘beyond the mountains’; *Levante*, from the east, the direction of the rising Sun; *Ostro*, from the south; *Ponente*, the west wind, from the direction of the setting Sun. This map shows an island named Caloiero, with a boat hoist designed to raise and lower the boat of the monks who lived on the high, rocky clifftop. While the woodcut maps in Sonetti's work are relatively simple, many of them have been hand-coloured and often have place names added by hand. Sonetti's work, one of the earliest volumes to include printed maps, was the first to be described as based on contemporary observations.

Imagines coeli septentrionales cum duodecim imaginibus zodiaci.

*Imagines coeli septentrionales*, Albrecht Dürer, Johannes Stabius and Conrad Heinfogel, 1515

This woodcut of the northern celestial hemisphere is, together with its southern hemisphere counterpart, the earliest known star chart printed in Europe. It was produced collaboratively. From the sheet showing the southern hemisphere we learn that Conrad Heinfogel (active 1515) plotted the constellations, Johann Stabius (1450–1522) worked on the projection and Albrecht Dürer (1471–1528) designed the figures of the constellations and produced the woodcut. Named in the corners of this chart are those on whose astronomical work Heinfogel and his colleagues had relied: Aratus of Soli, Claudius Ptolemy, Marcus Manilius and Abd al-Rahman al-Sufi. Dürer famously transformed the art of woodcutting, insisting that the cutters who worked under him reproduced every line on a drawing rather than simplifying what they cut, producing far more graphically sophisticated woodcut prints than had been made before.[56] That skill is most visible here in the delicacy of the constellation images, text and numbered stars and is a demonstration of the potential for including fine and elegant detail in woodcut maps. Dürer's star charts were hugely influential and used as the basis for several later celestial maps including the globe by Dutch cartographer Gemma Frisius.

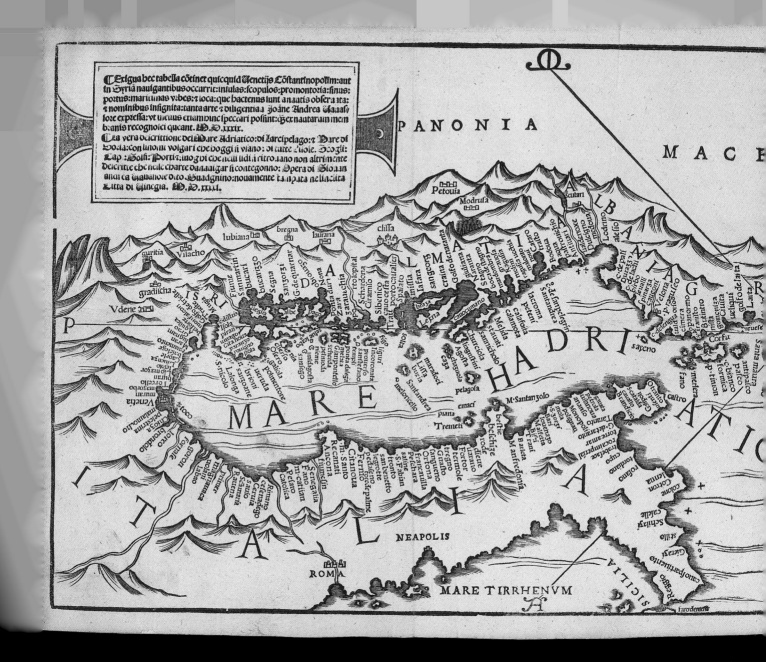

Drawing on the portolan tradition, with its place names perpendicular to the coast, scoop-shaped bays and directions of the winds, this is the earliest known printed map apparently intended for use at sea. Made by Venetian printmaker Giovanni Andrea Vavassore (1518–72), it provides details of the maritime route from Venice to Constantinople. Vavassore made it clear that this was a representation of navigational knowledge, describing on the map itself how it had been 'drawn with such care and skill by Giovanni Andrea Vavassore that the observer may see clearly here all that he previously knew from sailors' own

drawings'.[57] This, a second edition dating to 1541, is believed to be the only surviving copy; the first was published in 1539, indicating that there must have been a market for maps like this. Updating a woodcut, however, is complicated. While copper plates can be smoothed out by hammering and then re-engraved, the only thing you can do to amend a woodblock is remove more of the raised printing surface. Anything else requires a more elaborate fix. When Vavassore published his second edition, the new cartouche, in Italian rather than the Latin of the first edition,

was likely cut on a wooden plug, which would then have been inserted into the original block.[58] Integrating such a quantity of text into a woodcut map like this was unusual. Plugs made from cast metal would have been used to add the lettering on many woodcut maps of the period, because of the difficulty of cutting the rounded shapes popular in the Renaissance.[59] The fact that on this example the cartouche, with relatively extensive, close text, was itself cut from wood indicates the skill of those working in Vavassore's workshop.

*La vera descrittione del Mare Adriatico*, Giovanni Andrea Vavassore, 1541

**Y** IS FOR **YEAR**

The date a map was made is useful information for both practical users and historians. For people using maps for navigation, observation or planning, understanding whether a map might still bear resemblance to what it describes is key to assessing its reliability. When studying maps, a date provides vital context for the world in which a particular object was made and used. This also works the other way round: maps can be dated through the details they contain, sometimes revealing the inscribed date of making to be incorrect. Many maps, particularly those intended for practical purpose, contain multiple dates, indicating new revisions and editions and thus the reliability of the document. The date of making is not the only way that maps mark the passage of time. Since celestial observations have to be adjusted according to the date, understanding the course of a year became increasingly important for calculating location.

Most celestial globes go out of date, not because of the discovery of new stars, or the better fixing of star positions, but because of the very movement of the Earth relative to the cosmos. Due to a phenomenon known as the precession of the equinoxes (a steady change in the rotational axis of the Earth), the position of the stars in relation to a cartographic grid of the equator, the tropics and colures (great circles that intersect at right angles at the celestial poles) shifts over time. This means that a celestial globe is only accurate for a specific epoch and that an undated celestial globe can therefore be approximately dated according to its star positions, particularly as makers tended to use those relevant to their own times. While the epoch depicted on a celestial globe is not the only thing that can be used to establish its date, it does provide part of the picture. This Islamic celestial globe does not have the name of a maker engraved on it. However, by comparing the star positions with those listed by Ptolemy in 137 AD, the date of the globe has been established as 1596, with an uncertainty of 12 years.[60]

**Celestial table globe, unknown maker, around 1596**

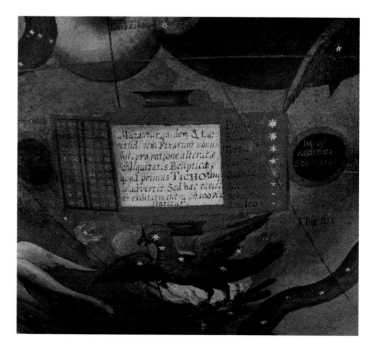

Terrestrial table globe,
unknown maker after Richard
Cushee, after 1752

This globe is a careful copy of one signed by Richard Cushee and
dated 1731. Scrutinising the horizontal ring that surrounds the
object shows that it was made at least 21 years later – and after
Cushee's death in 1733. In the sixteenth century, 'calendar drift',
or the way in which the calendar was increasingly out of sync with
the solar year, became intolerable to the Roman Catholic Church
because of the way it interfered with the timing of Easter. In the
Julian calendar, which had been in use since 46 BCE, a year was
calculated as exactly 365.25 days, when in fact it is fractionally
less. As a result, over the course of centuries, the dates of the
equinoxes shifted. While many Catholic countries adopted the new
Gregorian calendar in the 1580s, resetting the day of the month and
slightly reducing the number of leap years that occurred, Protestant
Britain only did so in 1752. That year, 2 September was followed
immediately by 14 September. Consequently, the way dates lined
up with the seasons and the divisions of the zodiac were altered
by 11 days, such that they would remain consistent over time. The
horizon rings of globes typically include a range of information, most
frequently the zodiac, the seasons and a calendar of months and
days. The particular alignment of these features – for instance, the

dates of solstices and equinoxes – on a horizon ring shows which
calendar has been used. This English globe uses the Gregorian
calendar, indicating that it was constructed after 1752, unlike the
original from which it was so diligently copied.

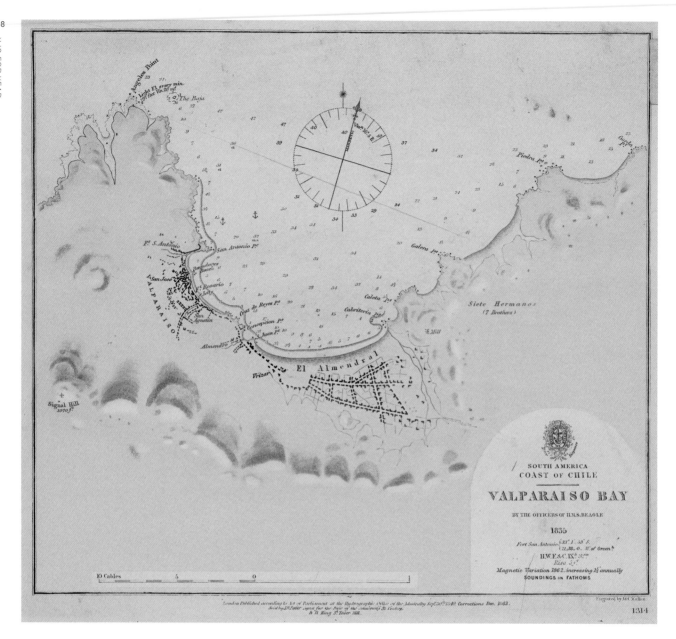

Admiralty charts include a lot of dates – of the survey, publication, corrections and editions. The survey work for this chart of Valparaiso Bay, Chile, was completed in 1835 by the officers of the Royal Navy ship, HMS *Beagle*. The chart itself was published in 1840, after an extensive process of drafting undertaken in London, and a corrected version was issued in 1863. The multiple dates draw attention to the intensive regime of revision and correction that was developed by the Hydrographic Office in the 1830s, but even they do not tell the whole story. Minor corrections, not yet incorporated into updated editions, were advertised in so-called 'Notices to Mariners', indicating that owners should correct their charts by hand or paste small, printed patches over an outdated section of coastline. As Hydrographer Francis Beaufort wrote, 'every day, intelligence arrives of the discovery of new rocks or sholas, or more correct limits being assigned to those already known. Improved means of observation are constantly yielding more accurate positions' and the rapid development of maritime infrastructure meant that details of lighthouses or the geography of ports was changing too.[61] Such revisions required navigators to pay close attention to the dates on charts, as a measure of how reliable they might be. Hydrographic work, then, is always in progress, never finished, however final a printed document may at first appear.

*South America, coast of Chile – Valparaiso Bay*,
Hydrographic Office, 1863 (first published 1840)

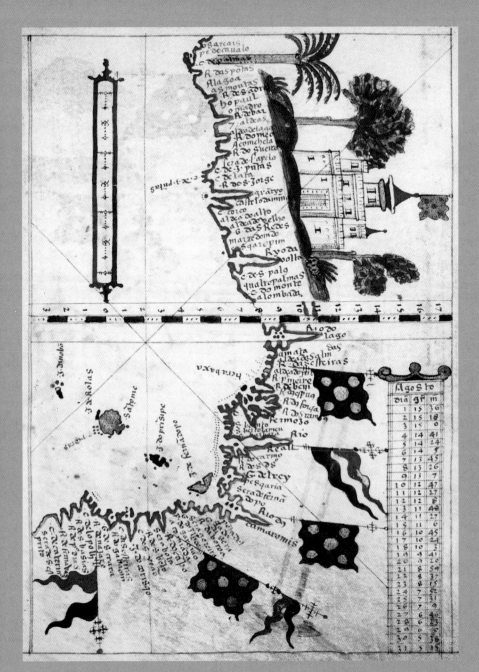

Map of West Africa in a volume
of manuscript maps, unknown
maker, around 1550

This small volume of manuscript maps uses a calendar to highlight the links between astronomical observation and geographical knowledge that were at the heart of Portuguese imperial expansion in the fifteenth and sixteenth centuries. On each map, there is a calendar for either one or two months. For every day of the year, these three-columned calendars – declination tables – show the angle between two imaginary lines: the plane of the Equator and a line going from the centre of Earth to the centre of the Sun. The value changes daily. At the solstices, the angle is at its largest: 23°28' north to 23°28' south in December. At the equinoxes, in March and September, it is at zero. Towards the end of the fifteenth century this information was crucial to Portuguese navigators, as they increasingly sailed south of the Equator and were no longer able to use the Pole Star to establish how far north or south they were. As a result, they began to measure latitude by the midday altitude of the Sun and observations then had to be interpreted using a table of solar declination like this.[62] This map of West Africa shows the location of the Portuguese forts from where hundreds of thousands of enslaved Africans were taken to Portugal and Brazil during the sixteenth century. While the volume contains maps far too small to be of navigational value, it makes the value of astronomy to geographical knowledge, and specifically Portuguese expansion overseas, abundantly clear.

STRAITS

DAN **Z** ABLE HE

60°

IS FOR

**ZONE**

FAREWELL

DEEP

50°

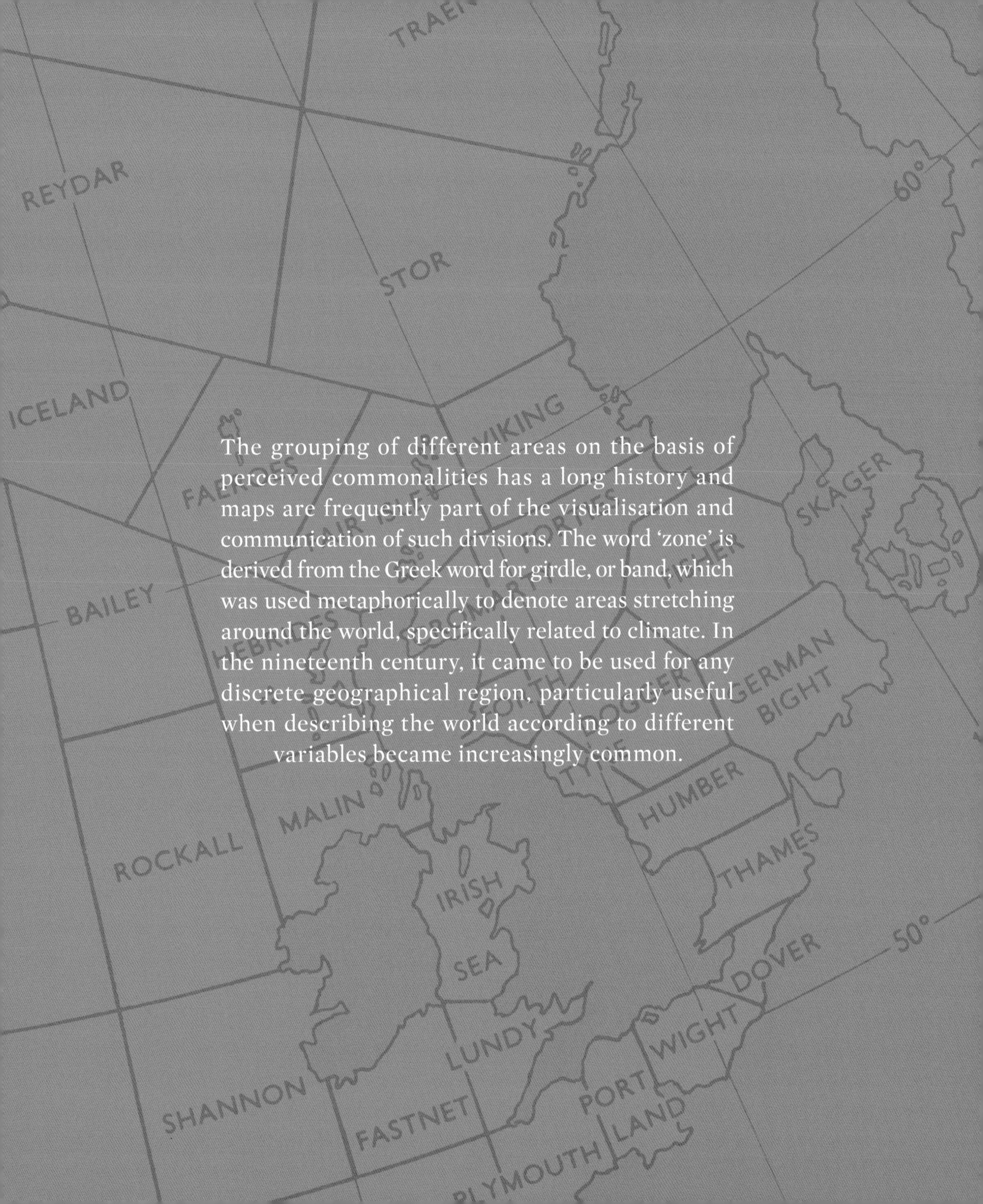

The grouping of different areas on the basis of perceived commonalities has a long history and maps are frequently part of the visualisation and communication of such divisions. The word 'zone' is derived from the Greek word for girdle, or band, which was used metaphorically to denote areas stretching around the world, specifically related to climate. In the nineteenth century, it came to be used for any discrete geographical region, particularly useful when describing the world according to different variables became increasingly common.

Zonal map, *Macrobii Aurelii Theodosii
... in somnium Scipionis*, published by
Johannes Soteris, 1527

Dividing the world into five zones, or *clima*, this map illustrates the theories of fifth-century Roman philosopher Macrobius, whose geographical work was hugely influential in medieval Europe. It was made for a sixteenth-century audience fascinated with classical learning, even as understanding of world geography had shifted enormously. The five zones, originally outlined by Greek philosopher Aristotle, were used to divide the world according to latitude. At the top and bottom of the map is the word *frigida*, which denotes the two frigid or cold zones closest to the poles. *Perusta* describes the torrid, hot, zone around the equator. In between two landmasses,

northern and southern continents that balance each other, is a great ocean, *alveus oceani*. To the south is the *temperata antipodum nobis incognita*, 'the southern temperate zone, to us unknown'. The northern temperate zone contains landmasses with more recognisable names – *italia*, *aphrica* among them. *Britania* sits at the edge of the northern frigid zone. Accompanying these geographical divisions were ideas about climate determinism – that is, how the nature of people living in particular places was shaped by the local climate. These persisted and intensified into racist theories which remained current into the twentieth century, even as other Macrobian ideas fell out of fashion.

*Great Britain – Fleet Forecast Areas*, Hydrographic Department, 1956

Somehow, the names of the fleet forecast areas around the British Isles have become bound up with a sense of national identity. It is an admittedly unusual outcome for a weather forecast. This map was published in 1956, at a time when the Shipping Forecast, which started as maritime weather warnings transmitted by telegraph in 1861 and was broadcast by the BBC from the 1920s, did not yet have its iconic status and was a crucial source of weather information for those at sea. It represents changes to the forecast areas which would, some 50 years later, become part of a heated debate. In 1955, meteorologists from the countries around the North Sea met to discuss marine weather forecasts and storm warnings. To improve the service, the large area 'Dogger' was divided in two and 'Fisher' introduced, and a portion of the region known as 'Forties' was split off to create 'Viking'. Danish and German meteorologists also suggested that the area 'Heligoland' be renamed 'German Bight' (the name they already used), as part of a process of international coordination that would help avoid confusion, especially important for a service concerned with the safety of life at sea. By the 2000s, the Shipping Forecast, with its mysterious, sometimes ominous phrasing and slow, rhythmic pace, had acquired a cultural status of its own, even as it was no longer the main source of weather information for shipping. When, therefore, a similarly pragmatic proposal was made to change the name of 'Finisterre' to 'FitzRoy' (thereby distinguishing it from an identically labelled Spanish region), it was met with vehement public opposition. The matter was viewed by some as being bound up with a politics of English identity in relation to continental Europe and, specifically, the European Union. In this context, the (in its day entirely uncontroversial) renaming of 'Heligoland' was resurrected as an earlier, lost battle for English exceptionalism.[64]

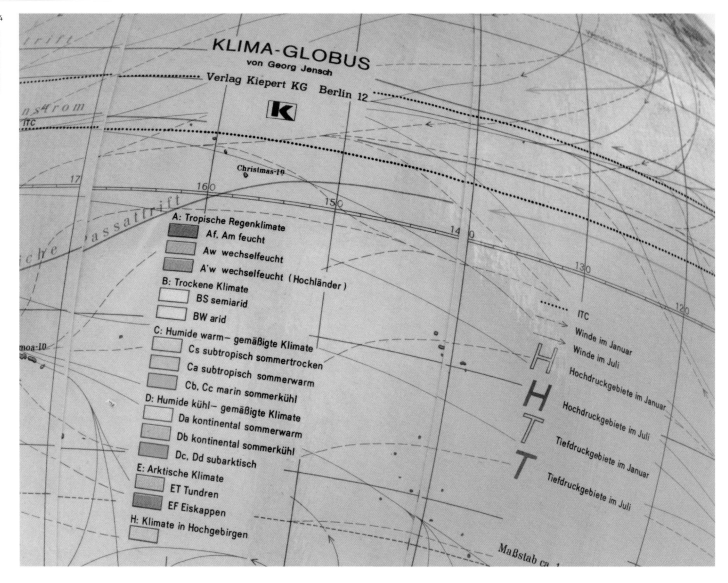

*Klima-Globus*, Georg Jensch, 1969

Designed for use in the classroom this climate globe by German geographer Georg Jensch shows the world divided according to the variables of vegetation, temperature and precipitation. The system it employs is known as the Köppen–Trewartha climate classification system, named after Wladimir Köppen, who developed the first system of global climate classification in 1900, and Glenn Trewartha, who modified it in the late 1960s. On this globe, tropical humid climates are coloured pink; dry climates yellow; subtropical regions are green; temperate climates blue; boreal areas purple; and polar climates white. Subtypes within these categories, which represent differences in either precipitation or temperature, are indicated by variations in tone. Today, in the context of accelerated human-induced climate change, scientists study earlier climate maps to understand how regional climates are being affected. The Köppen–Trewartha system remains one of the most established ways of classifying climatic zones and has been used to demonstrate the shrinking of the tundra, as well as significant changes in the global tropics.[63] Now 50 years old, this object is already a depiction of a global climate past.

# THE WORLD
## TIME ZONE CHART

### TIME NOW IN USE IN VARIOUS COUNTRIES ETC.

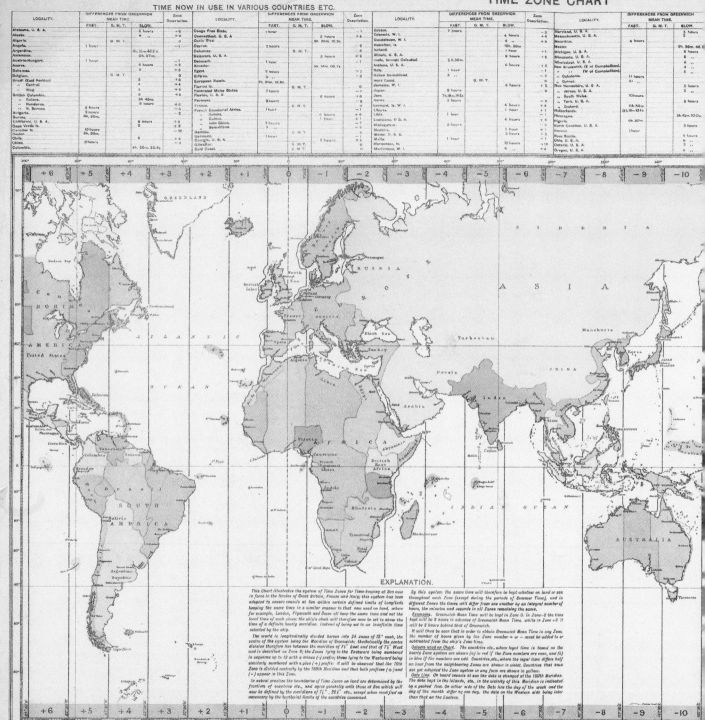

EXPLANATION.

This Chart illustrates the system of Time Zones for Time-keeping at Sea now in force in the Navies of Great Britain, France and Italy; this system has been adopted to ensure vessels at Sea within certain defined limits of Longitude keeping the same time in a similar manner to that now used on land, where for example, London, Plymouth and Dover all keep the same time and not the local time of each place; the ship's clock will therefore now be set to show the time of a definite hourly meridian, instead of being set to an indefinite time selected by the ship.

The world is longitudinally divided hereon into 24 zones of 15° each, the centre of the system being the Meridian of Greenwich; theoretically the centre division therefore lies between the meridian of 7½° East and that of 7½° West and is described as Zone 0; the Zones lying to the Eastward being numbered in sequence up to 12 with a minus (−) prefix; those lying to the Westward being similarly numbered with a plus (+) prefix; it will be observed that the 12th Zone is divided centrally by the 180th Meridian and that both prefixes (+) and (−) appear in this Zone.

In actual practice the boundaries of Time Zones on land are determined by the frontiers of countries etc., and agree generally with those at Sea which will now be defined by the meridians of 7½°, 22½° etc., except when modified as necessary by the territorial limits of the countries concerned.

By this system the same time will therefore be kept whether on land or sea throughout each Zone (except during the periods of Summer Time), and in different Zones the times will differ from one another by an integral number of hours, the minutes and seconds in all Zones remaining the same.

Examples. Greenwich Mean Time will be kept in Zone 0. In Zone −8 the time kept will be 8 hours in advance of Greenwich Mean Time, while in Zone +8 it will be 8 hours behind that of Greenwich.

It will thus be seen that in order to obtain Greenwich Mean Time in any Zone, the number of hours given by the Zone number + or − must be added to or subtracted from the ship's Zone time.

Colours used on Chart. The countries etc., where legal time is based on the hourly Zone system are shewn (a) in red if the Zone numbers are even, and (b) in blue if the numbers are odd. Countries, etc., where the legal time differs half an hour from the neighbouring Zones are shewn in violet. Countries that have not yet adopted the Zone system in any form are shewn in yellow.

Date Line. On board vessels at sea the date is changed at the 180th Meridian. The date kept in the Islands, etc., in the vicinity of this Meridian is indicated by a pecked line. On either side of the Date line the day of the week and the day of the month differ by one day, the date on the Western side being later than that on the Eastern.

London. Published at the Admiralty 21st March 1918, under the Superintendence of Rear Admiral J.F Parry, C.B. Hydrographer

In the decades around 1900, significant steps were made towards the standardisation of time, both nationally and internationally. Greenwich Mean Time (GMT) was legally adopted across Great Britain in 1880, the USA started using defined time zones across the country from 1883, and in 1891 the French authorities established Legal (Paris) Time. It took a little longer for time at sea to be regulated in this way, with ships adopting the relevant national time when they arrived in port but using apparent time (determined locally by the Sun) when on the move. As the *Geographical Journal* reported in 1918 this meant 'a ship was a law unto itself … there was no certainty that the time of a message despatched from the ship or of an entry in the ship's log could be translated into Greenwich mean time'.[65] It was in 1917 that Joseph Renaud, Director of the French Hydrographic Service of the Navy, had proposed a system of time zones to be used at sea. There were 24 zones, each an hour different to the one next to it, with GMT as the standard from which hours would either be added (moving east) or subtracted (moving west). While the hours would change, the minutes would always stay the same, so 0926 would become 1026 or 0826, depending on the direction travelled. Such an approach, which made time changes substantially easier, had not previously been used at sea. France and Italy quickly adopted the plan, with the British Admiralty following suit in 1919. This explanatory chart was made and issued as a result, indicating not only time zones at sea, but also relative times on land. New editions of this chart, a reference document that also symbolises an age of faster communication and places increased emphasis on standardisation, are still published today.

*The World – Time Zone Chart*,
Hydrographic Department, 1919

# NOTES

## INTRODUCTION

1   John Brian Harley and David Woodward, 'Preface' in John Brian Harley and David Woodward (eds), *The History of Cartography, Volume 1: Cartography in Prehistoric, Ancient, and Medieval Europe and the Mediterranean* (Chicago: University of Chicago Press, 1987), p. xvi.

2   David Woodward, 'Cartography in the Renaissance: Continuity and Change', in D. Woodward (ed.), *The History of Cartography, Volume 3: Cartography in the European Renaissance* (Chicago: University of Chicago Press, 2007), p. 20.

3   Geoffrey Callender to James Caird, 28 May 1935, NMM NMM5, Box 37.

## A IS FOR ATLAS

1   Samuel Pepys, quoted in Coolie Verner, 'John Seller and the Chart Trade', in Norman J. W. Thrower (ed.), *The Compleat Plattmaker: Essays on Chart, Map and Globe Making in England* (Berkeley: University of California Press, 1978), p. 140.

2   John Patrick Montaño, *The Roots of English Colonialism in Ireland* (Cambridge: Cambridge University Press, 2012).

3   Julie Chun Kim, 'The Caribs of St Vincent and Indigenous Resistance during the Age of Revolutions', *Early American Studies* 11:1 (2013), pp. 117–32; Bernard Marshall, 'The Black Caribs – Native Resistance to British Penetration Into the Windward Side of St Vincent 1763–1773', *Caribbean Quarterly* 19:4 (1973), pp. 4–19.

4   Lorenzo Pezzani and Charles Heller, 'Forensic Oceanography' https://forensic-architecture.org/subdomain/forensic-oceanography, accessed 06/12/2021.

5   'TH' in Petruccio Ubaldini, *A discourse concerning the Spanishe Fleete...*, translated by Augustine Ryther (London, 1590), n.p.

6   Augustine Ryther, 'To the Reader', ibid., n.p.

7   Benjamin Scott, *Practical Hints to Unpractised Lecturers to the Working Classes*, 4th ed. (London: Working Men's Education Union, 1858), pp. 7–8.

8   *Tee Emm*, December 1941, pp. 18–19.

9   Elly Dekker, *Globes at Greenwich* (Oxford: Oxford University Press, 1999), p. 371.

10  'Photolithography', *Scientific American* 8:11 (1863), p. 170.

11  Meg Roland, 'Facsimile Editions: Gesture and Projection', *Textual Cultures* 6:2 (2011), p. 55.

12  Emilie Savage Smith, *Islamicate Celestial Globes: Their Construction, History and Use* (Washington DC: Smithsonian Institution Press, 1985), pp. 43–44, 239.

13  Quoted in Christopher Lane, 'George Pocock, nineteenth-century maker of inflatable globes', *IMCOS Journal* 164 (2021), p. 14.

14  Martin White, *Sailing Directions for the English Channel* (London: Hydrographic Office, 1835), p. 1.

15  Thomas Hurd, 'Remarks and Observations respecting the Bermudas explanatory of their survey', 30 September 1801, reproduced in full in W. Powell Jones, 'Fortifying the Bermudas in 1801', *Huntington Library Quarterly* 4:4 (1941), p. 488.

16  Kevin Dawson, 'Enslaved Ship Pilots in the Age of Revolutions: Challenging Notions of Race and Slavery between the Boundaries of Land and Sea', *Journal of Social History* 47:1 (2013), p. 83.

17  C.G. Pitcairn Jones and J. Munday, 'The Surveyor's Curse', *Mariner's Mirror* 49:1 (1963), p. 53.

18  H.L.G. Bencker, 'Report Concerning the Third Edition of the General Bathymetric Chart of the Ocean', *International Hydrographic Review* 30:1 (1953), p. 86.

19  Quoted in Rainier III, Prince of Monaco, 'Foreword', *The History of GEBCO, 1903–2003* (Lemmer: GITC, 2003), n.p.

20  Elly Dekker, 'Globes in Renaissance Europe', in David Woodward (ed.), *The History of Cartography, Volume 3, Part 1: Cartography in the European Renaissance* (Chicago: University of Chicago Press, 2007), p. 154.

21  Alexander Pratt, 'The Occultation Machine of HM Nautical Almanac Office', *Journal of the British Astronomical Association* 124:1 (2014), pp. 12–21.

22  Francis Beaufort to Daniel Augustus Beaufort, 14 August 1814. Huntington Library MSS FB 372.

23  G. Earnest, *Two Years Adrift: the Story of a Rolling Stone* (Brighton: A M Robinson, 1870), p. 111.

24  Beatrice Grimshaw, *In the Strange South Seas* (London: Hutchinson & Co, 1907), p. 224.

25  'Fish: From the Sea to the Table' (White Fish Authority, n.d.), p. 1, 8.

26  Martin Wilcox, '"A record of abortive enquiries and empty of achievement?": the White Fish Authority 1951–1981', *Journal for Maritime Research* (2020), pp. 1–26.

27  Edward Evans MP (Lowestoft), 'White Fish Authority (Levy)' HC Deb 30 October 1956, *Hansard* vol. 558, p. 1401.

28  J. Goldsmith, *An Easy Grammar of Geography*, 40th edition (London: Phillips, 1811), p. iv.

29  Joseph Genz, 'Resolving Ambivalence in Marshallese Navigation: Relearning, Reinterpreting, and Reviving the "Stick Chart" Wave Models', *Structure and Dynamics: eJournal of Anthropological*

*& Related Sciences* 9 (1), 2016. https://escholarship.org/uc/item/43h1d0d7 (accessed 3/3/2021).

30  Translation by Michael Goeje, quoted in Nadja Danilenko, *Picturing the Islamic World: The Story of al-Istakhri's Book of Routes and Realms* (Leiden: Brill, 2021), p. 61.

31  Hugh Percival Wilkins, 'Notes on Lunar Drawing', *The Moon* 3:2 (1954), p. 43.

32  Horatio Nelson quoted in the *Hampshire Advertiser*, 18 June 1836.

33  *The Navigator*, Sept–Oct 1942, p. 171

34  G.S. Ritchie, *As It Was: Highlights of Hydrographic History from The Old Hydrographer's Column 'Hydro International'* (Lemmer, The Netherlands: GITC, 2003), p. 102.

35  *Yorkshire Post and Leeds Intelligencer*, 25 August 1945.

36  Walter Blanchard, 'The Genesis of the Decca Navigator System', *Journal of Navigation* 68:2 (2015), pp. 219–37.

37  'The Preparation and Use of Weather Maps by Mariners', World Meteorological Organization Technical Note no. 179, (Geneva: Secretariat of the World Meteorological Organization, 1966), p. vii.

38  Quoted in Melanie Vandenbrouck, 'From Service to Captivity: The Artist as Eyewitness', in Christine Riding (ed.), *Art and the War at Sea* (London: Lund Humphries, 2015), p. 114.

39  Eric Dolin, Leviathan: *The History of Whaling in America* (New York: W.W. Norton, 2007), p. 429.

40  Charles Piazzi Smyth, quoted in Katherine Anderson, *Predicting the Weather: Victorians and the Science of Meteorology* (Chicago: Chicago University Press, 2005), p. 189.

41  Anna Gambles, 'Free Trade and State Formation: The Political Economy of Fisheries Policy in Britain and the United Kingdom, circa 1780–1850', *Journal of British Studies* 39:3 (2000), p. 292.

42  Henry Brougham, quoted in Ian Barrow, 'India for the Working Classes: The Maps of the Society for the Diffusion of Useful Knowledge', *Modern Asian Studies* 38:3 (2004), p. 679.

43  Reports from Committees of the House of Commons (London, 1803), vol. 14, p. 278.

44  Marie Gillespie et al., 'Mapping Refugee Media Journeys: Smartphones and Social Media Networks', Open University/France Medias Monde Research Report, October 2016.

45  Chet van Duzer, *Sea Monsters on Medieval and Renaissance Maps* (London: The British Library, 2013), p. 103.

46  Robert Louis Stevenson, *Essays in the Art of Writing* (London: Chatto and Windus, 1905), p. 128.

47  Peter Barber and Rudolf Schmidt, 'Beyond Geography: Globes on medals 1440–1998', *Der Globusfreund* 47/48 (1999), p. 60.

48  Walter Ghim, quoted in Mike Zuber, 'The Armchair Discovery of the Unknown Southern Continent: Gerardus Mercator,

Philosophical Pretensions and a Competitive Trade', *Early Science and Medicine* 16:6 (2011), p. 5 15

49  'The Madeline Rock', *The Nautical Magazine* (December 1839), p. 811.

50  Tristan Stein, 'Tangier in the Restoration Empire', *The Historical Journal* 54:4 (2011), p. 989.

51  John Serres, quoted in Mike Barritt, *The Eyes of the Admiralty* (London: National Maritime Museum, 2008), p. 111.

52  *The British Critic* (London: F. and C. Rivington, 1801), vol. 18, p. 533.

53  Quoted in Vladimiro Valerio, 'Cartography in the Kingdom of Naples during the Early Modern Period', in David Woodward (ed.), *The History of Cartography, Volume 3, Part 1: Cartography in the European Renaissance* (Chicago: University of Chicago Press, 2007), p. 965, note 114

54  Ronald Doel, Tanya Levin and Mason Marker, 'Extending modern cartography to the ocean depths: military patronage, Cold War priorities, and the Heezen-Tharp mapping project, 1952–1959', *Journal of Historical Geography* 32 (2006), pp. 613–14.

55  Marie Tharp, 'Connect the Dots: Mapping the Seafloor and Discovering the Mid-ocean Ridge', in Laurence Lipsett (ed.), *Lamont-Doherty Earth Observatory of Columbia: Twelve Perspectives on the First Fifty Years 1949–1999*, reproduced at https://www.whoi.edu/news-insights/content/marie-tharp/ (accessed 1 September 2021).

56  Anthony Griffiths, *Prints and Printmaking* (London: British Museum, 1980), p. 12.

57  Translation by Natalie Lussey, 'Giovanni Andrea Vavassore and the Business of Print in Early Modern Venice' (PhD thesis, University of Edinburgh, 2016), p. 155.

58  Ibid.

59  Ibid., p. 157.

60  Elly Dekker, *Globes at Greenwich* (Oxford: Oxford University Press, 1997), p. 188.

61  Francis Beaufort to John Barrow, 20 December 1836, UKHO MB2, p. 345.

62  Eric Ash, 'Navigation Techniques and Practice in the Renaissance', in David Woodward (ed.), *The History of Cartography, Volume 3, Part 1: Cartography in the European Renaissance* (Chicago: University of Chicago Press, 2007), p. 518.

63  Michal Belda, Eva Holtanová, Tomáś Halenka and Jaroslava Kalvová, 'Climate Classification Revisted', *Climate Research* 59:1 (2014), pp. 1–13.

64  Victoria Carolan, 'The Shipping Forecast and English national identity', *Journal for Maritime Research* 13:2 (2011), pp. 105–07.

65  'Standard Time at Sea', *The Geographical Journal* 51:2 (1918), p. 97.

# FURTHER READING

## THE HISTORY OF CARTOGRAPHY PROJECT

As a comprehensive starting point for learning more about map history, readers will find plenty of interest in the volumes of the History of Cartography Project, which contain chapters on specific topics written by leading experts. The individual chapters of Volumes 1, 2, 3 and 6 are freely available online: https://press.uchicago.edu/books/HOC/index.html

J.B. Harley and David Woodward (eds), *The History of Cartography, Volume 1: Cartography in Prehistoric, Ancient and Medieval Europe and the Mediterranean* (Chicago: University of Chicago Press, 1987)

J.B. Harley and David Woodward (eds), *The History of Cartography, Volume 2, Book 1: Cartography in the Traditional Islamic and South Asian Societies* (Chicago: University of Chicago Press, 1992)

J.B. Harley and David Woodward (eds), *The History of Cartography, Volume 2, Book 2: Cartography in the Traditional East and Southeast Asian Societies* (Chicago: University of Chicago Press, 1994)

David Woodward and G. Malcolm Lewis (eds), *The History of Cartography, Volume 2, Book 3: Cartography in the Traditional African, American, Arctic, Australian, and Pacific Societies* (Chicago: University of Chicago Press, 1998)

David Woodward (ed.), *The History of Cartography, Volume 3, Parts 1 and 2: Cartography in the European Renaissance* (Chicago: University of Chicago Press, 2007)

Matthew Edney and Mary Pedley (eds), *The History of Cartography, Volume 4: Cartography in the European Enlightenment* (Chicago: University of Chicago Press, 2019)

Mark Monmonier (ed.), *The History of Cartography, Volume 6: Cartography in the Twentieth Century* (Chicago: University of Chicago Press, 2015)

## MAPS, POWER AND POLITICS

Peter Barber and Tom Harper, *Magnificent Maps: Power, Propaganda and Art* (London: British Library, 2010)

Jerry Brotton, *A History of the World in Twelve Maps* (London: Penguin, 2013)

Denis Cosgrove (ed.), *Mappings* (London: Reaktion, 1999)

John Brian Harley, *The New Nature of Maps: Essays in the History of Cartography* (Baltimore: John Hopkins University Press, 2002)

## MAPS, NAVIGATION AND EMPIRE

David Buisseret, *The Mapmakers' Quest: Depicting New Worlds in Renaissance Europe* (Oxford: Oxford University Press, 2003)

Surekha Davies, *Renaissance Ethnography and the Invention of the Human* (Cambridge: Cambridge University Press, 2016)

Jordana Dym, *Mapping Travel: The Origins and Conventions of Western Journey Maps* (Leiden: Brill, 2021)

Matthew H. Edney, *Mapping an Empire: The Geographical Construction of British India, 1765–1843* (Chicago: University of Chicago Press, 1997)

Katherine Parker and Barry Ruderman, *Historical Sea Charts: Visions and Voyages through the Ages* (Milan: White Star Publishers, 2021)

William Rankin, *After the Map: Cartography, Navigation, and the Transformation of Territory in the Twentieth Century* (Chicago: University of Chicago Press, 2016)

## MAPS, MAKING AND LEARNING

Will C. van den Hoonard, *Map Worlds: A History of Women in Cartography* (Waterloo: Wilfrid Laurier University Press, 2013)

Sumathi Ramaswamy, *Terrestrial Lessons: The Conquest of the World as Globe* (Chicago: University of Chicago Press, 2017)

Sylvia Sumira, *The Art and History of Globes* (London: British Library, 2014)

Judith A. Tyner, *Stitching the World: Embroidered Maps and Women's Geographical Education* (London: Ashgate, 2015)

David Woodward (ed.), *Five Centuries of Map Printing* (Chicago: University of Chicago Press, 1975)

# ACKNOWLEDGEMENTS

Books, like maps, are always the work of multiple hands. In a volume in which images play such an important part, first thanks must go to my colleagues in the Royal Museums Greenwich Photo Studio, Josh Akin, Charlotte Kite, Sophie Rogers, Sam Rowland and David Westwood. For all their work compiling images and seeking permissions to use them, I am extremely grateful to Rochelle Bisson and Louise Jarrold. For the care and movement of the objects themselves, many thanks to conservators Emmanuelle Largeteau and Paul Cook, Art and Object Handlers Roger Fell, Rob Hickerton, Dan Jackson and Ana Maria Lima Dimitrijevic, Store Managers Sarah Kmosena, Louise Bascombe and Debbie Williams, and Librarians Gareth Bellis and Stawell Heard. My colleagues in the Curatorial team have been hugely supportive as sounding boards and critics, and particular thanks must go to Robert Blyth, Lucy Dale, Louise Devoy and Aaron Jaffer, as well as to Simon Stephens for his endoscopy skills. Beyond the museum, the book has benefited from conversations with Richard Dunn, Matthew Edney, Katie Parker and Simon Schaffer. The exceptional eye of designer Ocky Murray has ensured that this volume is beautiful and does justice to the varied objects it describes. Sarah Connelly and then Kathleen Bloomfield have, with immense editorial expertise and good humour, guided the project from an idea into a manuscript into this book, and it has been an absolute pleasure to work with both of them. Finally, thanks must go to friends and family for their continued enthusiasm (you know who you are), and to Hobnob, who slept by my side while most of this text was written, and with great enthusiasm dragged me out for walks.

# PICTURE CREDITS

The publisher would like to thank the copyright holders for granting permission to reproduce the images illustrated. Every attempt has been made to trace accurate ownership of copyrighted images in this book. Any errors or omissions will be corrected in subsequent editions provided notification is sent to the publisher.

All the objects shown in this publication are from the collection of the National Maritime Museum and the object ID numbers are listed below. Unless otherwise stated, all images are © National Maritime Museum, Greenwich, London. Further information about the Museum and its collection can be found at rmg.co.uk.

pp. 6 and 7 ZBA5460
p. 8 G201:1/51 © National Maritime Museum, Greenwich, London. Caird Collection
p. 9 PBC3995(3) © National Maritime Museum, Greenwich, London. Caird Collection
p. 10 (left) GREN1C/5(A) © National Maritime Museum, Greenwich, London. Caird Collection
p. 10 (right, top and bottom) GLB0017 © National Maritime Museum, Greenwich, London. Caird Collection
p. 11 G230:1/4 © National Maritime Museum, Greenwich, London. Caird Collection
p. 13 P/36(3-4) © National Maritime Museum, Greenwich, London. Caird Collection
p. 14 G201:1/16
p. 15 (top) G213:2/4 © National Maritime Museum, Greenwich, London. Caird Collection
p. 15 (bottom) G201:1/29 © Crown copyright. Photo © National Maritime Museum, Greenwich, London
pp. 16–17 G235:3/39 © Internews. Photo © National Maritime Museum, Greenwich, London
p. 20 (top) PBD7640(3)
p. 20 (bottom) PBD7640(23)
pp. 20–1 PBD7640(1) © National Maritime Museum, Greenwich, London. Caird Collection
p. 22 PBD5253/1-12

pp. 22–3 PBD5253/1 © National Maritime Museum, Greenwich, London. Caird Collection
pp. 24–5 PBE6857 © National Maritime Museum, Greenwich, London. Macpherson Collection
pp. 26–7 PBD8034 © National Maritime Museum, Greenwich, London. Caird Collection
pp. 30–1 P/49(20) © National Maritime Museum, Greenwich, London. Caird Collection
pp. 32 (detail) and 33 GREN80A/1 © National Maritime Museum, Greenwich, London. Caird Collection
pp. 34–5 (detail) and 36 PBH8042(23) © National Maritime Museum, Greenwich, London. Caird Collection
p. 37 ZBA9423 © Forensic Oceanography; licensed to the National Maritime Museum as part of the acquisition. Acquired with Art Fund support
pp. 40–1 G230:1/7 © National Maritime Museum, Greenwich, London. Caird Collection
pp. 42–3 PBD8529(2)
p. 44 AAA4825 © National Maritime Museum, Greenwich, London. Caird Fund
p. 45 ZBA8883 © Mark Wallinger. Photo © National Maritime Museum, Greenwich, London
pp. 48–9 G294:1/3
p. 50 G290:1/2 and G292:1/2
p. 50–1 G296:1/1
p. 52–3 G218:9/6 © Crown copyright. Photo © National Maritime Museum, Greenwich, London
pp. 54 and 56–7 (detail) ZBA4552
p. 55 G298:1/6 © Crown copyright. Photo © National Maritime Museum, Greenwich, London
p. 60 PBD8264(12) © National Maritime Museum, Greenwich, London. Macpherson Collection
p. 61 PBD8186/2 © National Maritime Museum, Greenwich, London. Macpherson Collection
pp. 62 (detail) and 63 PBB6192
p. 64 (centre) ZBA8754 and ZBA8751

Reproduced with the permission of Roy Cooney's family. Photo © National Maritime Museum, Greenwich, London
pp. 64–5 G298:1/5 © Crown copyright. Photo © National Maritime Museum, Greenwich, London
p. 68 GLB0177 © National Maritime Museum, Greenwich, London. Caird Collection
p. 69 GLB0036 © National Maritime Museum, Greenwich, London. Caird Collection
pp. 70–1 G301:1/70(1)
pp. 72–3 GLB0253 © National Maritime Museum, Greenwich, London. Caird Collection
pp. 76 and 77 (detail) GLB0175 © National Maritime Museum, Greenwich, London. Caird Collection
pp. 78 and 79 (detail) GLB0086 © National Maritime Museum, Greenwich, London. Caird Collection
pp. 80 and 81 (detail) GLB0123 © National Maritime Museum, Greenwich, London. Caird Collection
pp. 82 and 83 (detail) GLB0230
p. 86 G218:6/12
p. 87 G214:5/2 © Crown copyright. Photo © National Maritime Museum, Greenwich, London
pp. 88 (detail) and 89 G235:3/40
pp. 90–1 G266:5/1 © IHO-IOC GEBCO. Photo © National Maritime Museum, Greenwich, London
pp. 94 (detail) and 95 GLB0135 © National Maritime Museum, Greenwich, London. Caird Collection
p. 96 BGY/D/4/5
p. 97 GLB0053 © National Maritime Museum, Greenwich, London. Caird Collection
pp. 98 and 99 NAV1803
pp. 102–03 G201:1/43 © National Maritime Museum, Greenwich, London. Caird Collection
p. 104 PBD7680
p. 105 P/18(3) © National Maritime Museum, Greenwich, London. Caird Collection
pp. 106 (detail) and 107 PBC5347
pp. 110–11 G297:20/15

pp. 112 (detail) and 113 G235:8/5 © Crown copyright. Photo © National Maritime Museum, Greenwich, London

pp. 114 and 115 (detail) G267:23/3

pp. 116 and 117 (detail) G215:1/7(1) © White Fish Authority/Seafish. Photo © National Maritime Museum, Greenwich, London

pp. 120–1 TXT0046

pp. 122–3, p. 123 AAB0099

p. 124 GLB0187

p. 125 ZBA7550

pp. 128 (detail) and 129 P/3 © National Maritime Museum, Greenwich, London. Caird Collection

pp. 130–1, p. 132 In P/12 © National Maritime Museum, Greenwich, London. Caird Collection

p. 133 G218:6/21 © National Maritime Museum, Greenwich, London. Caird Collection

pp. 134 and 135 (detail) ZBA8769.9

pp. 138 and 139 (detail) PBD8166(1)

p. 140 MRY298:11/3

p. 141 G298:4/20 © Nautilus International and Hensoldt UK. Photo © National Maritime Museum, Greenwich, London

pp. 142 (detail) and 143 G223:4/27 © Crown copyright. Photo © National Maritime Museum, Greenwich, London

pp. 146 and 147 (details) GLB0157

pp. 148 and 149 PBG2055, figs A, F and G © National Maritime Museum, Greenwich, London. Airy Collection

pp. 150 (detail) and 151 G288:1/2(2) © Crown copyright. Photo © National Maritime Museum, Greenwich, London

pp. 152–3 G201:1/82 World Meteorological Organisation

pp. 156–7 PBD8170 © Crown copyright. Photo © National Maritime Museum, Greenwich, London

pp. 158 and 159 GLB0236

pp. 160 and 161 ZBA5274 (recto and verso) and ZBA5278 (recto and verso) © National Maritime Museum, Greenwich, London. Presented by the artist's family, 2012.

pp. 162 and 163 (detail) ZBA8787 © Art Refuge

UK. Photo © National Maritime Museum, Greenwich, London. Acquired with Art Fund support.

pp. 166 and 167 (detail) STK201:7/6 © National Maritime Museum, Greenwich, London. Caird Fund.

p. 168 G298:2/5(1) © Crown copyright. Photo © National Maritime Museum, Greenwich, London

p. 169 G298:2/5(2) © Crown copyright. Photo © National Maritime Museum, Greenwich, London

pp. 170 (detail) and 171 GREN2B/2 © National Maritime Museum, Greenwich, London. Caird Collection

pp. 172 (detail) and 173 PBP3510/2

pp. 176 and 177 (details) G297:25/3

pp. 178 and 179 (detail) GLB0034 © National Maritime Museum, Greenwich, London. Caird Collection

p. 180 GRB218:9/41(3)

p. 181 ZBA8919 Courtesy of National Maritime Museum collection/Orphan work

pp. 184–5 PBD7617(131)

pp. 186 and 187 G290:1/8E

pp. 188–9 In PBC5309 © National Maritime Museum, Greenwich, London. Caird Collection

pp. 190–1 PBE9994

p. 194 PBP1512

pp. 195, 196–7 DUF292:2/2

p. 198 MEC0027

p. 199 P/33(19) © National Maritime Museum, Greenwich, London. Caird Collection

p. 202 P/6(1) © National Maritime Museum, Greenwich, London. Caird Collection

p. 203 GLB0178

pp. 204–05 G214:11/11 © Crown copyright. Photo © National Maritime Museum, Greenwich, London

pp. 206 (detail) and 207 G224:4/8 © Crown copyright. Photo © National Maritime Museum, Greenwich, London

p. 210 PBD8026(63)

p. 211 P/43(9)

pp. 212–13 PBD8139 (plate 8)

pp. 214 (detail) and 215 PBD8518 (plate 25)

pp. 218–19 G201:1/53 © National Maritime Museum, Greenwich, London. Caird Collection

pp. 220 (detail) and 221 G201:1/13 © National Maritime Museum, Greenwich, London. Caird Collection

pp. 222–3 G201:1/45

pp. 224–5 G201:1/57 Courtesy of Marie Tharp Maps, LLC. Copyright © Fiona Schiano-Yacopino

pp. 228–9 PBE5109 © National Maritime Museum, Greenwich, London. Caird Collection

p. 230 PBC5214 (folio 34) © National Maritime Museum, Greenwich, London. Macpherson Collection

p. 231 G200:1/6A © National Maritime Museum, Greenwich, London. Caird Collection

pp. 232–3 G235:1/3

p. 236 GLB0106 © National Maritime Museum, Greenwich, London. Caird Collection

p. 237 GLB0115 © National Maritime Museum, Greenwich, London, Caird Collection

p. 238 G279:4/39 © Crown copyright. Photo © National Maritime Museum, Greenwich, London

p. 239 P/14(6R) © National Maritime Museum, Greenwich, London, Caird Collection

p. 242 PBE9632(110) © National Maritime Museum, Greenwich, London, Caird Collection

p. 243 G213:3/25 © Crown copyright. Photo © National Maritime Museum, Greenwich, London

pp. 244 (detail) and 245 GLB0204

pp. 246–7 G201:1/84 © Crown copyright. Photo © National Maritime Museum, Greenwich, London

Images from the collection of Royal Museums Greenwich, the world's greatest source of historical maritime images, can be licensed for a range of media, including book publishing and digital content, from images.rmg.co.uk, and a large selection are also available from prints.rmg.co.uk as high-quality prints, canvases and cards.

# INDEX

Published in 2022 by the National Maritime Museum,
Park Row, Greenwich, London, SE10 9NF

ISBN: 978-1-906367-93-0

At the heart of the UNESCO World Heritage Site of Maritime
Greenwich are the four world-class attractions of Royal Museums
Greenwich – the National Maritime Museum, the Royal Observatory,
the Queen's House and *Cutty Sark*.

www.rmg.co.uk

Designed by Ocky Murray

Colour reproduction by Altaimage, London

Printed and bound in Belgium by Graphius

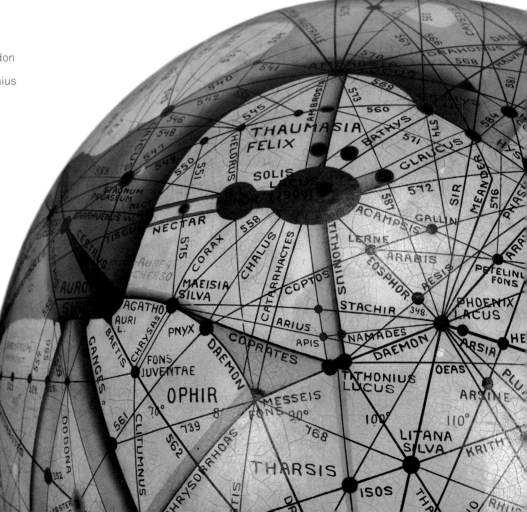